UNVEILING

www.royalcollins.com

UNVEILING

A Deep Dive into the Architecture Powering Our Connected Future

By Tan Shiyong, Ni Hui, Zhang Wanqiang,
Wu Xiaobo, and Hu Huadong

Books Beyond Boundaries

ROYAL COLLINS

Unveiling 5G:
A Deep Dive into the Architecture Powering Our Connected Future

By Tan Shiyong, Ni Hui, Zhang Wanqiang, Wu Xiaobo, and Hu Huadong

First published in 2024 by Royal Collins Publishing Group Inc.
Groupe Publication Royal Collins Inc.
BKM Royalcollins Publishers Private Limited

Headquarters: 550-555 boul. René-Lévesque O Montréal (Québec) H2Z1B1 Canada
India office: 805 Hemkunt House, 8th Floor, Rajendra Place, New Delhi 110008

B&R Book Program

Original Edition © Publishing House of Electronics Industry
This English edition is authorized by Publishing House of Electronics Industry.

ISBN: 978-1-4878-1166-2

To find out more about our publications, please visit www.royalcollins.com.

Contents

Foreword   *vii*

Preface   *xi*

Chapter 1   Overview   **1**

1.1   The fundamental concept of wireless network architecture and core
       network   1
1.2   The evolution of wireless network architecture   2
1.3   Deployment mode of the 5G network architecture   8
1.4   Commercial value of the 5G network architecture   10
1.5   Standard protocol map of the 5G network architecture   11
       References   12

**Chapter 2   The Driving Force behind Designing a New 5G Network
               Architecture**   **13**

2.1   Driven by services   14
2.2   Operation deployment   18
2.3   Design principles of the 5G network architecture   24
       References   27

Chapter 3   Key Features of 5G Network Architecture   **29**

3.1   Overview of 5G network architecture   29
3.2   Basic features of the 5G network architecture   36
3.3   Service-based architecture   111
3.4   Network slicing   119
3.5   Edge computing   129

3.6   Network intelligence   138
3.7   IoT   174
3.8   Telematics   194
3.9   Ultra-Reliability and Low Latency Communication   211
3.10   Convergence of fixed and mobile networks   218
3.11   Security   250
3.12   Vertical industry-oriented features   262
       References   271

Chapter 4   An Outlook on the Future Evolution of 5G Network Architecture   273

4.1   3GPP version release pace   273
4.2   Continuous evolution of the 5G network architecture   275
       References   280

Abbreviations   *281*

Index   *287*

Foreword

4G mobile communication networks have contributed to the explosive growth of the mobile Internet since it became commercially available on a global scale in 2010. A variety of Internet applications and innovations that rely on smartphones and high-speed cellular connections have emerged, changing lifestyles on a profound level. After years of development, 4G mobile communications have achieved phenomenal commercial success and become an indispensable communication infrastructure in daily life. By the end of 2019, the total number of mobile users in China had exceeded 1.6 billion, with a penetration rate of over 100%, and the total number of nationwide 4G base stations exceeded 4.3 million. Mobile terminals have become the main way for people to access the Internet. The leapfrog development of mobile communication has also provided new opportunities for all industries to promote their development of information, as well as digital and mobile transformation. On this basis, with 5G networks moving from theory into reality, the mobile communication industry has started to evolve from 4G to 5G and is witnessing yet another new transformation.

With the popularization of mobile broadband, users' demand for a better mobile network service experience is also increasing. Mobile network service represented by 4K/8K HD video, VR/AR, real-time gaming, online shopping, and mobile payment requires rapid improvement of bandwidth, latency, and security accordingly. Based upon 4G, the 5G network needs to enhance communication bandwidth and reduce transmission latency to provide consumers with a better mobile broadband experience and offer a more solid connection foundation for the flourishing of the developing mobile Internet, thus giving rise to more new mobile services and service models.

In addition, the reach of 5G is extending to vertical industries. With the advancement of industry digitization, various industries (including ICT, finance, retail, automobile, factories, and agriculture) are undergoing continuous transformation. 5G networks shoulder the heavy responsibility of supporting the Internet of Things and digital industry transformation. 5G is generally expected to provide unified connectivity for all industries and support the rapid digital transformation and service innovation of the industry by building a more complete network architecture for public communications and vertical industries in general. In terms of network requirements, unlike mobile broadband services, many vertical industries have very demanding requirements for the performance of some communication networks (such as latency, reliability, and data security), posing new challenges to the design and deployment of the 5G networks.

At the same time, a series of ICT evolution technologies (represented by network virtualization, Cloud computing, software-defined networks, network slicing, edge computing, and artificial intelligence) are also maturing and forming initial deployments during the same period. 5G networks are facing a revolutionary change in their deployment and operation environment, with virtualization and distributed deployment of network elements becoming a trend. 5G network architecture needs to be designed in a way that comprehensively supports a variety of differentiated services, such as enhanced mobile broadband, ultra-reliability, low-latency wireless communications, massive IoT connectivity, and other typical scenarios. It is also necessary to consider how to meet the network requirements of the new deployment environment and how to support the joint deployment of heterogeneous systems (e.g., 4G/5G) and reduce operational complexity, which makes it difficult to conceive a design for the network architecture.

With the joint efforts of the industry, the first standardized version of 5G—Rel-15—was released in June 2018, providing the support capability for the 5G basic services. Rel-16 is scheduled to be set in March 2020, enhancing the support capability for various vertical industries. 5G network architecture uses new design concepts such as modularization, service orientation, control and forwarding separation, and access independence. This makes the 5G network architecture simpler, more flexible, and easier to scale compared to the 4G network architecture, so that it can be adapted to a wide variety of service needs and deployment scenarios through unified network architecture. For more complex deployment scenarios such as Cloud-based deployment and edge computing, the 5G network architecture is designed with new control and forwarding mechanisms such as element registration discovery, network slicing, and user-plane flexible routing to meet the evolutionary trends of future mobile networks.

In 2013, the Ministry of Industry and Information Technology, the National Development and Reform Commission, and the Ministry of Science and Technology jointly promoted the establishment of the IMT-2020 (5G) Promotion Group, which aims to promote domestic 5G network research. It also systematically arranges for the uplink and downlink industry chains (including operators, equipment vendors, terminal manufacturers, and chip vendors) to participate in and promote next-generation mobile communications research and standardization and has incubated a series of fruitful achievements. Furthermore, Huawei cooperates with operators and industrial partners on 5G standardization. Based on its experience in designing and operating support for mobile communication networks, it always designs common system network architectures and interfaces around service requirements and the goals of flexible deployment and simplified operation, making active contributions to the 5G standardization of 3GPP and ITU. Frank Mademann, the technical expert of its network architecture, chaired the 3GPP SA2 Network Architecture Working Group from the beginning of the 5G network architecture research project to the completion of the Rel 15 standardization, ensuring its successful completion with his professional technical level and efficient organization. Through the joint efforts of the industry chain, the standardization of the 5G network architecture has made rapid progress, and the simultaneous support of non-independent and independent 5G grouping architectures was completed in the first version of 3GPP for 5G.

In June 2019, China released its 5G operating license, officially launching the commercial deployment of the 5G networks. Several operators were planning to commercially deploy 5G Core Networks in 2020 to support the launch of independent 5G group network services. This will be the world's largest deployment of the 5G networks and will lay a solid foundation for China's industrial innovation and continued leadership in the 5G phase, providing additional lessons in the practical use of 5G for countries around the world.

Focusing on the evolution of the 5G network architecture, this book explains the service, deployment drivers, and overall design principles of the 5G network architecture from basic concepts. It focuses on the implementation principles and standardization schemes of each fundamental and enhanced feature of the 5G network architecture and provides an outlook on its future evolution.

The lead author, Tan Shiyong, is the head of the Network Technology Group of China's IMT-2020 (5G) Promotion Group. The other authors have been involved in 3GPP-related research and standardization projects for many years and have a deep understanding of the standardization of the 5G network architecture, from the identification of initial-use cases and requirements to basic solution identification and standardization, and subsequent evolution trends. The material in this book includes the latest standard protocols and technical contributions from various standard organizations, including 3GPP, as well as the authors' summary of and discussions on core issues in the standardization process. The book can be used as a reference for communications research practitioners to understand the ins and outs of the 5G network architecture standardization and development trends, as technical guidance for developers of communications products, and as a reference text for the selection of features for the commercial deployment of core 5G networks.

WAN LEI
Fellow, Huawei Technologies Co.
Director, Wireless Standards and Patents Department, Huawei
3GPP Standards Expert
IMT-2020 Promotion Group Expert
January 2020

Preface

4G mobile communication technology has been a major commercial success. It announced the arrival of the mobile Internet era and changed daily life forever. At the same time, new demands are emerging, and mobile communication networks are once again requiring upgrades. Data traffic is growing rapidly, the massive Internet of Things (IoT) is taking shape, and new services are putting forward higher requirements for the timeliness and reliability of the network. 5G technology inherits the historic mission of changing society. Identifying how to meet the varying needs of mobile operators and thousands of industries and how to build a smart world where everything is connected, poses a huge challenge to the design of the 5G network architecture.

In July 2012, the International Telecommunication Union's Radiocommunication Sector launched a study on the future of wireless communications for 2020 and beyond. Between 2013 and 2015, many countries and regions (including China, Europe, South Korea, Japan, and the United States) set up organizations to promote 5G. The mobile communication industry, research institutions, and universities became deeply involved in research into 5G technologies, which laid a solid foundation for the development of international 5G standards.

In February 2013, China established the IMT-2020 (5G) Promotion Group with the joint promotion of the Ministry of Industry and Information Technology, the National Development and Reform Commission, and the Ministry of Science and Technology, overseen by the China Academy of Information and Communications Technology. This is the basic platform on which China can aggregate the strengths of the industry, academia, research, and applications in the field of mobile communication to promote research into 5G mobile communication technology and carry out international exchanges and cooperation. I was honored to be the leader of the Network Technology Group under the Promotion Group and was responsible for arranging and coordinating the technical research and standard promotion and testing the specification development of domestic 5G network architecture. In the years that followed, along with many experts in 5G network architecture research and standards, I saw China achieve "5G leadership" after "2G following, 3G breakthrough, and 4G synchronization."

3GPP—the core organization that develops international standards for 5G technology—began performing research related to 5G network architecture in 2015 and released the first international 5G standard version in 2018. This was the first milestone in 5G standardization; the 5G non-stand-alone deployment standard was set in March 2018, and the 5G stand-alone deployment version was set in June 2018. By June 2019, 3GPP had completed the development of the second version of the network architecture (Stage 2) standard and will set the second 5G international standard version by early 2020. The non-independent deployment standard has been commercialized in several countries, and the 5G independent deployment standard is expected to enter daily use in 2020. IMT-2020 is the official name of the International Telecommunication Union's Radio Communication Group for 5G. Perhaps this numerical coincidence foreshadows the fact that 5G technology will augur a new era and build a smart society where everything is connected.

Returning to the theme of this book, the term "architecture" is plainly understood as the division and interrelationships among the elements of a system. In the case of wireless networks, architecture is the division of the functions of each network element in a wireless communication system and the definition of the interfaces among each element. Unlike wireless air interface technology, the advantages and disadvantages of wireless network architecture are difficult to assess quantitatively through technical means such as simulation, and there are many dimensions to assess, including scalability, flexibility, complexity, openness, reliability, security, ease of deployment, and ease of maintenance. In addition, thousands of network devices have been deployed in the existing network, and these devices cannot all be upgraded or retired at once. Therefore, the definition of each generation of network architecture needs to consider the depreciation of existing network devices. This is a huge burden, and the design of the new-generation network architecture must strike a balance between revolutionary disruptive innovation and conservative smooth evolution, so as to guarantee the commercial success of the entire industry.

Based on the latest international standards for the 5G network architecture (Rel-15 and Rel-16), this book provides a comprehensive and systematic introduction to the design ideas and technical solutions of the 5G network architecture. It includes a review of the history of the evolution of mobile network architecture, an introduction to the driving forces defining 5G network architecture, various key technologies of 5G architecture, and an estimate of the direction of future evolution.

The book explains all aspects of 5G architecture in a systematic way in four chapters. Chapter 1 explains the concepts, evolutionary paths, deployment models, service values, and standard protocol maps of wireless network architectures. Chapter 2 explores the driving forces and basic design principles for the design of the 5G network architecture. Chapter 3 provides an in-depth discussion of the overall design of the 5G network architecture and the main technical features and solutions. In addition to basic features such as mobility management, session management, Quality of Service (QoS) and policy control, voice and interoperability with 4G, 5G network features are also analyzed, such as service architecture design, network slicing, edge computing, network intelligence, IoT, Telematics, ultra-reliability and low-latency communication, and fixed-mobile network convergence. In addition, chapter 3 introduces the capabilities of the 5G network architecture for security and vertical industry support. Chapter 4 provides an outlook on the subsequent evolution of the 5G network architecture. In the process of creating this book, we have tried to work within the limits of international standards. At the same time, we have tried to dissect the service and technical drivers behind the standards and integrate insights and reflections from the process of developing the 3GPP standards, allowing for a thorough understanding of the content.

This book is the result of collective wisdom and cannot be separated from the hard work of colleagues engaged in research on 5G network architecture standards for many years and the technical contributions of experts from the Network Technology Group of the IMT-2020 (5G) Promotion Group under the supervision of the Ministry of Industry and Information Technology and the China Academy of Information and Communication Research. I would like to dedicate this book to the research and standard experts who have contributed to 5G network architecture, to the IMT-2020 (5G) Promotion Group, and to the operators and the participating companies that support 5G.

All of the members of the writing team for this book come from the 5G network architecture standards research team of Huawei Technologies, and all have participated in the development of international standards for the 5G architecture. The Huawei Technologies 5G network architecture standards research team is a leader in this development and is the initiator and driver of important technologies such as service-oriented architecture, slicing, control-plane and user-plane separation, and network intelligence. The team members also include key personnel such as the chair of 3GPP SA2 and the chair of the 3GPP CT Plenary. In addition to the bylined authors, other contributors to this book include Li Meng, Zhu Hualin, Sun Haiyang, Zhou Runze, Huang Zhenglei, Cui Yang, and Li Huan. In addition, Wan Lei, Shen Zukang, Li Yan, Chen Zhongping, Yang Yanmei, Jin Weisheng, Wu Rong, Wang Yuan, Yu Fang, Ma Jingwang, Wu Yizhuang, Ying

Jiangwei, Li He, Hu Li, Zhu Fangyuan, Li Yongcui, Ge Cuili, Chong Weiwei, Xin Yang, and Xu Shengfeng reviewed and proofread this book, and provided valuable revisions, for which we express our most sincere gratitude.

Finally, we welcome readers and experts to point out any issues or mistakes in the text.

TAN SHIYONG

January 3, 2020

CHAPTER 1

Overview

1.1 The fundamental concept of wireless network architecture and core network

The mobile communication network is a complex system consisting of a terminal, an Access Network, and a Core Network, which are organically combined by network architecture—the major engineering drawings of this complicated architecture. The network architecture determines the functional module structure of the entire mobile communication system and the way to realize a variety of mobile communication services through the combination of different functional modules.

From an initial voice call to a full 5G service scenario, the services supported by mobile communication systems have been significantly enriched. Network architecture has evolved from the initial Global System for Mobile Communications (GSM) Network to the current 5G service-oriented and smart network. Moreover, the function of the network has proceeded from the initial Circuit Switched Domain to the Packet Switched Domain bearing management and strategy control, which is a huge leap.

The Core Network is the brain and the hinge of mobile communication. It provides the most fundamental but critical services, such as user subscription management, access safety authentication and authorization, user location tracking, mobility management, connection establishment and release, service quality and strategy control, and accounting. It also provides various mobile communication services for the terminal in conjunction with the base stations. The Core Network is responsible for mobile communication exchange services, routing, interconnection, and service control, so as to guarantee the success of end-to-end mobile communication services and realize the global roaming among different operators.

The wide application of mobile Internet and the full-service 5G scenario also promote the increasingly important role played by the Core Network in the mobile communication system.

1.2 The evolution of wireless network architecture

As the wireless network developed from 2G and 3G to 4G and 5G, both service and technology served as the major forces, driving the evolution of end-to-end mobile networks to a targeted structure and eventually the formation of a smart and fully connected pipeline based on a Cloud-deployed Core Network. In the process of this evolution, the wireless network diverged into Circuit Switched Domain and Packet Switched Domain, leading to a major change in the services they support:

(1) Circuit Switched (CS) Domain: also known as CS Domain. The 2G/3G Circuit Switched Domain has evolved into a full IP Packet Switched domain, forming a unified voice network based on the IP Multimedia Subsystem (IMS).
(2) Packet Switched (PS) Domain: also known as PS Domain. Full IP, flat network architecture, separation of control and bearing, decoupling of network functions, and the formation of a service-oriented network function design.
(3) Service evolution: evolving from voice, SMS, and other basic capabilities to three service scenarios of full 5G connection.

1.2.1 2G

The initial 2G network could only provide basic services like voice and SMS with its CS Domain. For the demands of data transmission, a Core Network element (known as GPRS Core Network) was subsequently added to support PS data services. The 2G network architecture is shown in figure 1-1 [1].

The Mobile Switching Center (MSC) is used to connect the wireless system and fixed network and is the core of the CS Domain network. It provides an exchange function and is responsible for mobile user paging access, channel allocation, call connection, traffic control, billing, and base station management. It also provides an interface function to other functional entities and fixed networks of the system, including the Public Switched Telephone Network (PSTN), the Integrated Services Digital Network (ISDN), and the Public Data Network (PDN). As the core of the network, the MSC works with other network units to complete the functions of mobile user location registration, handoff and automatic roaming, legitimacy verification, and channel switching.

The Visitor Location Register (VLR) is a unique device in the CS Domain. It serves mobile users in the control area, stores information about the registered mobile users who enter the control area and provides the necessary data for these users to establish call connection. When the Mobile Station (MS) roams to a new VLR area, the VLR initiates location registration to its Home Location Register (HLR) and obtains necessary user data. When the MS roams outside of the control range, it will re-register in another VLR, and the original VLR will cancel the temporarily recorded mobile user data. Therefore, the VLR can be regarded as a dynamic user database.

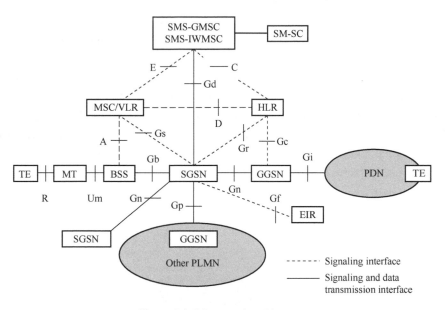

Figure 1-1 2G network architecture

The Serving GPRS Support Node (SGSN) is the core of PS Domain. It provides a connection between the Core Network and wireless access system BSS and has a connection interface for Gateway GPRS Support Node (GGSN)/MSC/HLR in the Core Network. The SGSN completes functions of packet data services such as mobility management and session management. It also manages the mobile and communication services of the MS in the mobile network and provides billing information. The GGSN is also a unique PS Domain device, which is the gateway that connects the GPRS network and external network. The GGSN provides packet routing and encapsulation between the mobile network and external data network. It is connected to the SGSN through the GN interface and the external data network (Internet/Intranet) through the GI interface.

1.2.2 3G

The Core Network of 3G is still based on that of GSM/GPRS, which is divided into the CS Domain and PS Domain. The Rel-99 version of the 3rd Generation Partnership Project (3GPP) introduces a new Wideband Code Division Multiple Access (WCDMA) wireless interface transmission technology in the wireless access part, but the Core Network has not been fundamentally changed. The CS Domain is based on the original GSM network, and the PS Domain is based on the original GPRS network. The 3GPP Rel-4 version implements the concept of soft switch in the CS Domain of the Core Network; that is, the traditional MSC is separated into Media Gateway (MGW) and MSC server and begins to move towards being complete IP integration. 3GPP Rel-5 introduces IMS into the Core Network to control the transmission of real-time and non-real-time multimedia services in the PS Domain. 3GPP Rel-6 version also introduces Multimedia Broadcast Multicast Service technology to support streaming media services such as mobile phones and TV more effectively. In the later stage of its evolution, designers of the 3G Core Network began to consider the efficiency of user side data processing. The 3GPP Rel-7 version introduces a direct tunnel from the Access Network base station to the GGSN. Since the SGSN link is omitted in the user plane path, the routing and transmission efficiency is improved. The 3G network architecture is shown in figure 1-2.

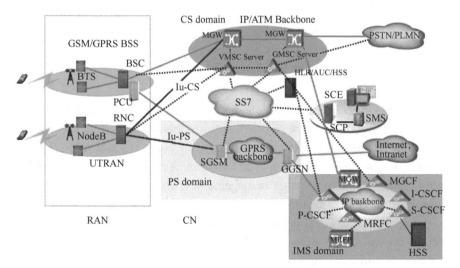

Figure 1-2 3G network architecture

1.2.3 4G

With the explosive growth of data services, the mobile network introduced Long Term Evolution (LTE) wireless access technology to support a higher data transmission rate, while the Core Network introduced an Evolved Packet Core (EPC), which no longer supports the CS Domain whose services are now provided by IMS or returned to the 2G/3G CS Domain. Compared with the traditional GPRS network, the functions of the 4G Core Network are significantly enhanced:

(1) It constructs full IP networks in an end-to-end way and adopts a flat architecture to reduce the network level. The wireless access part evolved from the Radio Network Controller (RNC) plus NodeB in the 3G era to an eNodeB node.
(2) It adopts Full PS Domain networking in support of IP broadband services with a high rate, low delay (24/7 online), and enhanced quality.
(3) It supports different access systems, including non-3GPP access modes, and supports users in roaming and switching between 3GPP and non-3GPP networks.
(4) It supports the selection of access system based on information including operator policies, user preferences, and Access Network conditions.
(5) It provides the capacity to coexist with traditional networks and transform from traditional networks to evolutionary one.
(6) It provides stronger security based on more effective encryption algorithms.

The 4G network architecture is shown in figure 1-3 [2]. The Mobility Management Entity (MME) is the core node of the control plane in the 4G Core Network. It is responsible for managing the mobility of the control plane, managing user context and mobile status, and allocating temporary identities to users.

The Serving Gateway (S-GW) is the user plane anchor point between different Access Networks in 3GPP and is responsible for the exchange of user plane data when users move between different access technologies. The S-GW terminates the base station user plane.

The Packet data network Gateway (P-GW) connects with the external Packet Data Network (PDN), which undertakes the gateway function of EPC and terminates the SGi interface connected with PDN. A terminal can access multiple PDNS through multiple P-GW at the same time. S-GW and P-GW are usually deployed together in one physical network element.

The Policy and Charging Rules Function (PCRF) fulfills the function of dynamic QoS policy control and flow-based charging control and in the meantime, provides authorization control based on users' signing information.

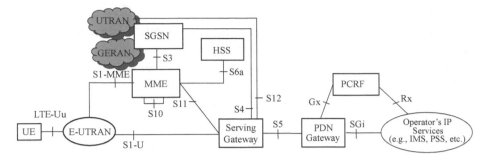

Figure 1-3 4G network structure

In the later stage of the 4G network architecture, Control and User Plane Separation (CUPS) was introduced to divide the gateway entity function into a control plane and a user plane. It supports both the centralized control plane and distributed user plane, so as to facilitate the independent upgrading of the network control plane of operators and the user plane. The only remaining function of the user side is data forwarding, which makes it more convenient for operators to achieve flexible deployment and lays a foundation for the evolution to the 5G network architecture. The network architecture of CUPS is shown in figure 1-4 [3].

Within the CUPS network architecture, the control plane sends other rules such as data forwarding and billing to the user plane, and the user plane carries out data forwarding, information statistics, and reporting based on these rules. Both the control plane and the user plane can allocate user plane tunnel information and cache downlink data in the idle state.

With the gradual enhancement of network capability, the interaction demand between the network and external applications is becoming stronger. The Rel-13 version introduces the capability-open architecture, which adds the Service Capability Exposure Function (SCEF) entity as a bridge connecting the internal entities of the 3GPP network and external applications and defines the corresponding South interface and North interface. It supports basic capability-opening functions, including

(1) event monitoring: the User Equipment (UE) accessibility, the UE location, loss of UE connection, roaming status of UE, number of UEs in a specific area, and downlink data transmission failure;
(2) reporting the overall network status: the congestion status of the network in a specific area);
(3) background data transmission;
(4) specific QoS data transmission;
(5) changing the third-party billing entity.

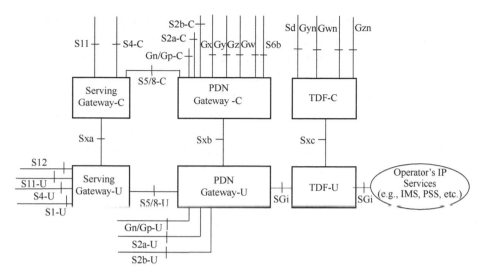

Figure 1-4 The CUPS network architecture

Capability-open architecture is shown in figure 1-5 [4].

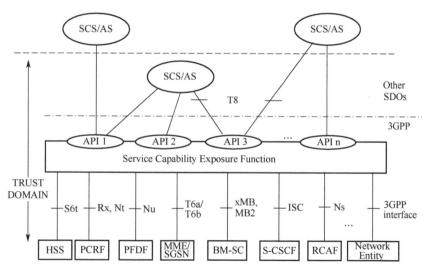

Figure 1-5 Capability-open architecture

1.2.4 5G

With the efforts of the entire industry, 5G network architecture has emerged in the face of the future full service and fully connected mobile service scenario. It introduces a large number of new technologies, driving the evolution of mobile network architecture in a flexible and smart direction.

Service architecture is the cornerstone of the 5G network architecture, which distinguishes itself from 2G, 3G, and 4G. The 5G network architecture decouples the traditional module functions and defines several Network Functions (NF) based on the open Application Programming Interface (API). Each network function is presented and called through the Network Function Service (NF service). An NF provides services to any other NF that allows the use of these services through a service-oriented interface. Within this design framework and principle, each network element of the Core Network control plane provides a service interface based on Hypertext Transfer Protocol (HTTP), and the elements communicate by calling each other's service interface. These service-invocation relationships are combined with each other through standardized timing and parameters, which eventually form the control process of various services in the 5G network.

The 5G service network architecture is shown in figure 1-6 [2].

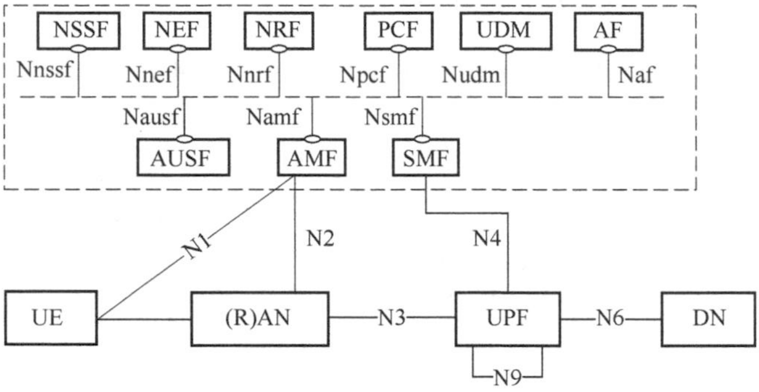

Figure 1-6 The structure of the 5G service network

1.3 Deployment mode of the 5G network architecture

The deployment mode of the 5G network architecture is shown in figure 1-7. Based on the 4G deployment scale of the current network and future evolution strategy of different operators, 5G supports a variety of deployment modes:

(1) Option 2 refers to the independent deployment of New Radio (NR) to access the 5G Core Network.

(2) Option 3 refers to the dual connection mode of accessing EPC with LTE, which is used as the anchor point of the control surface.

(3) Option 4 refers to the dual connection mode with NR, which is used as the anchor point to access the 5G Core Network.

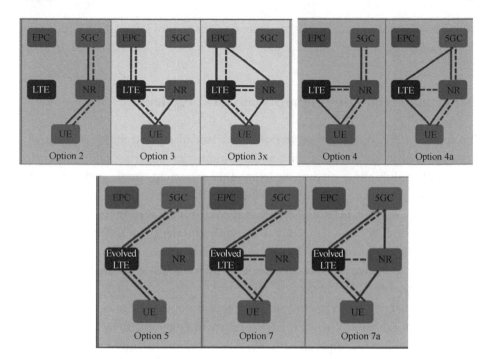

Figure 1-7 Deployment mode of the 5G network architecture

(4) Option 5 refers to accessing the 5G Core Network in the Evolved LTE independent deployment mode.

(5) Option 7 refers to adopting the Evolved LTE as the anchor point of the control plane to access the 5G Core Network in a dual connection mode.

Among the above five deployment options, Option 3 is proposed as certain operators are eager to apply 5G technologies for commercial use, which brings about the needs to upgrade the 4G EPC. Since the traditional EPC/4G Core Network is used for service control, and QoS and bearer management still adopt the 4G mode, it is considered a transitional solution.

Since Option 2, Option 4, Option 5, and Option 7 use the newly defined 5G Core Network, they can make full use of the advantages of service-oriented architecture and Cloud deployment, which is conducive to operators' increasing revenue and reducing expenditure in 5G network deployment and quickly applying it to new services. Option 2 uses a new air interface NR and a new 5G Core Network, which can make full use of the performance gain and economic benefits brought by the new technology. This will be the main 5G deployment mode of mainstream operators in the future. The numbering of the options above extends the serial numbers in the standard discussion; other options are not standardized due to the lack of support.

1.4 Commercial value of the 5G network architecture

From 4G to 5G, operators' services have expanded from traditional voice and data services to hundreds of thousands of industries, covering multiple service scenarios of all kinds, including smart medical treatment, smart grid, smart city, and smart car services. As a diversity of services has different requirements on the network, the wireless network has been driven from the traditional mode of "we will try our best" to a guaranteed network with experience of "differentiation and certainty." As a full-access and full-service enabling center, 5G Core Network has the ability to perceive the network, services and users, which is the key to realize "differentiation and certainty."

5G significantly improves the performance of mobile networks. The peak rate is more than ten times higher than 4G, up to 10 Gbps; the delay is reduced by more than 50 times compared with 4G, down to 1ms; the number of terminals supported by the cell has also been increased by two orders of magnitude. Based on the substantial improvement of the above basic capabilities, it makes the user experience of the original 4G service better and provides a stronger pipeline capacity to carry new services such as 4K/8K and Virtual Reality (VR) / Augmented Reality (AR).

According to the needs of future service development of the operators, the service-driven architecture defines an efficient, agile, and open network architecture to meet the challenges brought by future service uncertainty, so as to maximize the commercial value of the 5G network.

1. HIGH EFFICIENCY
(1) Independent lifecycle management: service autonomy, independent deployment, independent capacity expansion and reduction, and independent evolution.
(2) Function decoupling: network function decoupling and reconfiguration.

2. AGILITY
Flexible arrangement and free combination are carried out according to the needs of services, so as to speed up the online time of new services and shorten the Time to Market (TTM).

(1) Service interface: each NF only needs to define a unified interface.

(2) Service governance and orchestration: the fast-publishing mechanism per-
     fectly supports grayscale upgrading and grayscale publishing through the
     service registration and discovery mechanism provided by the network.
(3) Stateless design: processing and data are separated; policy implementation
     and data are separated.

3. OPENING UP
Helps operators expand new markets and develop new service models, evolving
from traditional MBB market to vertical industry market and from B2C to B2B.

(1) Open architecture for easy integration of third-party, customized services.
(2) Carries out the network capability openness through Network Exposure
     Function (NEF) and Application Function (AF) to help operators build a
     win-win ecosystem.

1.5 Standard protocol map of the 5G network architecture

The 5G network architecture protocol can be downloaded from the 3GPP official
website (https://www.3gpp.org/ftp/Specs/archive/). The main protocols of the 5G
network architecture are briefly explained below:

TS 23.501 [5]: an overview of the 5G network architecture, which explains the
network architecture (roaming and non-roaming), i.e., a summary of the network
functions and basic characteristics, the overall service framework, the service
discovery mechanism, and the service list.

TS 23.502 [6]: a detailed process description of basic characteristics of the 5G
network and a detailed description of the network element service.

TS 23.503 [7]: a detailed description of the policy control process.

TS 23.288 [8]: a smart network, which explains the intelligent architecture,
data collection process, collection feedback process, and definition of the Network
Data Analytics Function (NWDAF)'s network element service.

TS 23.316 [9]: 5G fixed-mobile convergence, which explains the fixed-mobile
convergence architecture, the scheme of a wireline terminal accessing the 5G
Core Network, and the scheme of supporting fixed broadband through a wireless
network.

TS 23.287 [10]: vehicle internet communication, which explains the architecture
of the Internet of Vehicles and the relevant processes of the PC5 interface and Uu
interface.

It is important to note that basic features such as mobility management,
session management, voice, interworking, slicing, and ultra-reliability, and low-

latency communication are described in TS 23.501/23.502. In addition, there are no separate protocols for IoT and Vertical LAN, which are also described in TS 23.501/23.502.

References

1. 3GPP. Technical Specification 23.060: General Packet Radio Service (GPRS); Service description; Stage 2 [Internet]. 2019 Mar [cited 2019 Dec 8]. Available from: http://www.3gpp.org/ftp/Specs/archive/23_series/23.060/

2. 3GPP. Technical Specification 23.401: General Packet Radio Service (GPRS) enhancements for Evolved Universal Terrestrial Radio Access Network (E-UTRAN) access [Internet]. 2019 Jun [cited 2019 Dec 8]. Available from: http://www.3gpp.org/ftp/Specs/archive/23_series/23.401/

3. 3GPP. Technical Specification 23.214: Architecture enhancements for control and user plane separation of EPC nodes; Stage 2 [Internet]. 2019 Jun [cited 2019 Dec 8]. Available from: http://www.3gpp.org/ftp/ Specs/archive/ 23_series/23.214/

4. 3GPP. Technical Specification 23.682: Architecture enhancements to facilitate communications with packet data networks and applications [Internet]. 2019 Jun [cited 2019 Dec 8]. Available from: http://www. 3gpp.org/ftp/Specs/archive/23_series/23.682/

5. 3GPP. Technical Specification 23.501: System Architecture for the 5G System; Stage 2 [Internet]. 2019 Jun [cited 2019 Dec 8]. Available from: http://www.3gpp.org/ftp/Specs/archive/23_series/23.501/

6. 3GPP. Technical Specification 23.502: Procedures for the 5G System; Stage 2 [Internet]. 2019 Jun [cited 2019 Dec 8]. Available from: http://www.3gpp.org/ftp/Specs/archive/23_series/23.502/

7. 3GPP. Technical Specification 23.503: Policy and Charging Control Framework for the 5G System; Stage 2 [Internet]. 2019 Jun [cited 2019 Dec 8]. Available from: http://www.3gpp.org/ftp/Specs/archive/23_series/23.503/

8. 3GPP. Technical Specification 23.288: Architecture enhancements for the 5G System (5GS) to support network data analytics services [Internet]. 2019 Jun [cited 2019 Dec 8]. Available from: http://www.3gpp. org/ftp/Specs/archive/23_series/23.288/

9. 3GPP. Technical Specification 23.316: Wireless and wireline convergence access support for the 5G System (5GS) [Internet]. 2019 Jun [cited 2019 Dec 8]. Available from: http://www.3gpp.org/ftp/Specs/ archive/23_series/ 23.316/

10. 3GPP. Technical Specification 23.287: Architecture enhancements for the 5G System (5GS) to support Vehicle-to-Everything (V2X) services [Internet]. 2019 Aug [cited 2019 Dec 8]. Available from: http://www. 3gpp.org/ftp/Specs/archive/23_series/23.287/

CHAPTER 2

The Driving Force behind Designing a New 5G Network Architecture

From the perspective of service needs, when the standard definition was first initiated, the goal of 4G network architecture was to meet people's needs for wireless broadband data communication. After that, applications such as support for IoT and telematics were introduced in the evolution of subsequent standard versions. The initial positioning of the 4G network is much lower than that of 5G—a form of people-oriented communication that is also used to solve various scenarios on the Internet of Things, such as wide coverage, large connections, ultra-reliability, and low delay. 5G is not only considered for wireless access but also for the scenario of integrating various wireline access and wireless fixed access, for the traditional network services of operators, and also to meet the differentiated needs of thousands of industries. The goal of 5G network architecture is to meet various communication scenarios such as person-to-person, person-to-object, and object-to-object, as well as to support the wireless communication needs of operators and vertical industries. Therefore, from the initial positioning, 5G is quite different from 4G. If 4G's positioning is to change daily life, then 5G's positioning is to change society as a whole, bringing the digital world into every household and organization and building an intelligent interconnected world.

At the same time, during more than a decade of 4G standard development and commercialization, various network-related technologies have also been making progress. Operators are eager to use these new technologies to simplify the deployment and operation of mobile communication networks. Virtualization and cloud technology have had very mature applications in the IT field. After years of exploration in the industry, they have come to be valued and accepted by the CT community. They have also been introduced to mobile communication networks and have begun to be applied in the commercial deployment of the 4G network. For example, typical products in the Huawei CloudEdge series simplify the deployment and operation of operator networks through virtualization and cloud technology. The idea of control, forwarding, and separation of a Software Defined Network (SDN) is also valued by 3GPP. It introduced CUPS into the 4G network architecture, which is a typical characteristic of 4.5G network architecture. With the introduction of the CUPS feature, the user side of the gateway is significantly simplified. It can be deployed to a location closer to the user on a large scale, so as to reduce data transmission delay on the user side, reduce the bandwidth demand of the return network, and pave the way for the actual deployment of mobile edge computing. In addition, Big Data and Artificial Intelligence Technology have also made breakthrough progress, entering people's lives from the laboratory. Identifying how to use these emerging technologies to improve the intelligence of the network, so that operators can deal more easily with the increasingly large and complex network has become a popular topic in the communication industry.

The difference between service requirements and deployment and operation requirements will inevitably lead to great changes in 5G architecture compared with 4G architecture. The progress and development of various related technologies over the past decade have provided a solid technical foundation for defining a new 5G network architecture.

2.1 Driven by services

2.1.1 Introduction
ITU divides 5G applications into three scenarios, namely Enhanced Mobile Broadband (eMBB), Ultra Reliable and Low Latency Communication (URLLC), and massive Internet of Things (mIoT). The 5G-oriented application scenario [1] is shown in figure 2-1. It is noteworthy that these three application scenarios are not completely orthogonal. For example, some applications require both high bandwidth and low delay.

4G changes lives, while 5G changes society. As the main infrastructure of the information age, the 5G network has thousands of different types of applications. Looking back on 4G, when the network architecture was first designed, it was difficult to imagine that there would be such a variety of applications on today's 4G network. In fact, before the real large-scale commercial deployment of the 5G network, it was difficult to predict how many different types of applications 5G network would spawn. Ten typical 5G applications are given in the "White Paper on the Top 10 Application Scenarios in the 5G Era" [2], including Cloud VR/AR, the Internet of Vehicles, smart manufacturing, smart energy, wireless medical treatment, wireless home entertainment, networked UAV, social network, personal AI assistance, and smart cities. In this book, only some of the application scenarios are used to illustrate the possible prospects of the 5G network. The following chapters quote some of the "White Paper on the Top 10 Application Scenarios in the 5G Era."

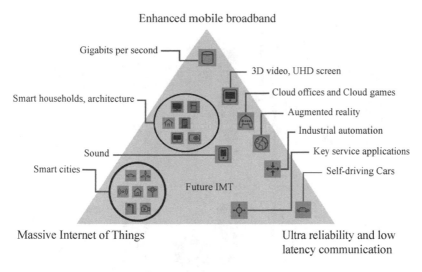

Figure 2-1 The 5G-oriented application scenario

2.1.2 Enhanced mobile broadband

eMBB refers to people-oriented mobile broadband applications. There are similar application scenarios in IoT and telematics, but only Cloud VR/AR is used as an example to illustrate in this part. Virtual Reality (VR) and Augmented Reality (AR) are revolutionary technologies that can completely subvert the traditional human-computer interaction mode. Occupying a huge market space, reform is reflected in the field of consumption and in many commercial and enterprise markets. ABI Research estimates that by 2025, the total AR and VR markets will reach $292 billion (with $151 billion in the AR market and $141 billion in the VR market).

VR/AR requires substantial transmission, storage, and computing functions. If these data- and computing-intensive tasks are transferred to the cloud, the data storage and high-speed computing capacity of the cloud server can significantly reduce the equipment cost and promote VR/AR to the market quickly. The demand for bandwidth is huge as high-quality VR/AR content processing moves to the cloud, and it may become the highest-traffic service in mobile networks. Even though the average throughput of existing 4G networks can reach 100 Mbps, some high-order VR/AR applications still require higher speed and lower latency. As can be seen in figure 2-2, real-time Computer-Generated Cloud rendering VR/AR requires a network delay of less than 10ms and a large bandwidth of up to 100 Mbps–9.4 Gbps.

	Stage 0/1		Stage 2	Stage 3/4
	PC VR	Mobile VR	Cloud Assisted VR	Cloud VR
VR application and technical features	Games, Modeling (Local render action and local loop)	360 video, education (Panoramic video download, local closed-loop action)	Immersive content, Interactive simulation, Visual design (Motion Cloud closure, FOV(+) video download)	High-quality real-time rendering and download of game and modeling experience (Motion Cloud closure, Cloud CG rendering, FOV(+) video download)
	2D AR		3D AR/Mixed Reality	Cloud MR
AR application and technology features	Operation simulation and guidance, gaming, telecommuting, retail, marketing visualization (Local overlay of image and text)		Spatially expanding holographic visualization, highly networked AR applications on public security (Image upload, corresponding multimedia information in the cloud)	Cloud-based mixed reality applications with increased user density and connectivity (Image upload, re-rendering in the Cloud)
Connectivity requirements	4G and Wi-Fi Streaming as content 20 Mbps + 50 ms latency requirement	Dominant Wi-Fi connectivity	4.5G Streaming as content 40 Mbps+20 ms latency requirement	5G Streaming as content 100 Mbps–9.4 Gbps+2–10 ms latency requirement

Figure 2-2 The five stages of Cloud VR/AR evolution

2.1.3 Ultra reliability and low latency communication

Compared with 4G, high reliability and low latency are significant improvements in the network communication performance of 5G. Only smart manufacturing is illustrated in this part. Innovation is part of the core competitiveness of the manufacturing industry. At present, the main development directions of the manufacturing industry include lean production and digitization, as well as workflow and production flexibility. In short, it is developing towards smart manufacturing. In the traditional mode, manufacturers rely on wireline technology to connect applications. In recent years, wireless solutions such as Wi-Fi and Bluetooth have been established in manufacturing workshops, but these solutions have limitations in bandwidth, reliability, and security. For example, synchronous real-time cooperative robots require a network delay of less than one millisecond, which is far from what 4G networks can achieve. At the 2017 Mobile World Congress (MWC), Huawei and KUKA demonstrated 5G cooperative robots with the two robots beating drums in a synchronous manner. The KUKA Innovation Laboratory reported that the network delay was as low as one millisecond, and the reliability was up to 99.999%.

According to statistics, by the end of 2017, there will be 18 million condition-monitoring connections in the world. By 2025, this number will rise to 88 million. The global shipment of industrial robots will also increase from 360,000 to 1.05 million. At present, the wireline network plays a leading role in a number of industrial IoT connections. However, the forecast shows that from 2022 to 2026, the Compound Annual Growth Rate of the 5G Industrial Internet of Things will reach 464%.

Smart manufacturing is a very promising application field for 5G. With the reduction of the time delay and the improvement of reliability, 5G will enter the diverse vertical industry market, altering thousands of industries and completely changing society.

2.1.4 Large-scale IoT applications

5G will continue 4G's support for the IoT and will significantly improve network coverage, terminal power consumption, and system cost. Only the smart city is used as an example in this part. Unlike the IoT in the 4G era, which focused on low-cost, small data and large connection application scenarios, 5G will be more oriented to IoT scenarios with high-bandwidth requirements, such as video surveillance—a very important element in smart cities.

Urban video surveillance is a highly valuable tool. It can help improve the security of the city and aid enterprises in improving their work efficiency significantly. Typical monitoring scenarios include (1) Busy public places (squares, activity centers, schools, hospitals), (2) Commercial areas (banks, shopping centers, squares), (3) Transportation centers (stations, wharves), (4) Major intersections, (5) Areas with high crime rates, (6) Institutions and residential areas, (7) Flood control (canals, rivers), (8) Key infrastructure (energy networks, telecommunication data centers, pump stations).

As cameras integrate increasingly powerful data collection and analysis functions, the demand for video surveillance continues to grow. At present, the cameras that dominate the market are 4m pixels, 6m pixels, and 8m pixels. 4K resolution surveillance cameras will go to the market in 2020, matching the commercial rhythm of 5G. The latest video surveillance camera has many enhanced features, such as high frame rate, ultra-high definition, and Wide Dynamic Range (WDR), which can produce images even in poor lighting conditions. These features will widen the application prospect of video surveillance and will generate more data traffic.

Currently, the UK has deployed six million cameras, and many other countries are stepping up the deployment of video surveillance equipment. It is estimated that by 2025, the value-added service revenue in the non-consumer video surveillance market will reach $21 billion.

2.1.5 Conclusion

According to the service requirements of various types of applications, ITU summarizes the key performance indicators of 5G, as shown in figure 2-3. Most network capabilities have been improved by one or even two orders of magnitude. Quantitative change leads to qualitative change, and the improvement of performance indicators has posed major challenges for the design of the network architecture. Traditional network architecture has had difficulty meeting the service needs of the new era. It is, therefore, necessary to design a new network architecture to truly lay the foundation for building an intelligent interconnected world.

2.2 Operation deployment

2.2.1 Introduction

The large-scale commercial use of mobile communication networks began with 2G and encompassed 3G and 4G till today. 2G defines the CS network architecture, which mainly solved the problem of voice communication, and later introduced the PS network architecture, i.e., GPRS, with data communication capabilities. 3G made a series of modifications to the CS network architecture, such as control and forwarding separation and strengthened data communication capabilities. 4G realized mobile broadband and gave birth to the mobile Internet era.

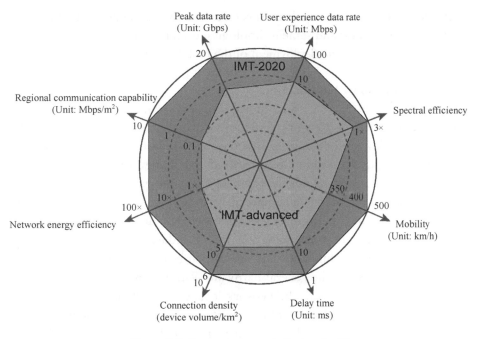

Figure 2-3 Key performance indicators for 5G

Due to the constraints of various factors such as network investment in profitability, the persistence of old terminals, and the limitations of spectrum usage, the development pattern of mobile communication networks is not a case of one generation completely replacing the previous one, but the long-term coexistence of multi-generation networks. Currently, many operators have 2G, 3G, and 4G network equipment in their networks at the same time. Each generation of the mobile communication network itself is a huge and highly complex system, containing dozens of network elements. The coexistence of multiple generations of networks, coupled with the fact that these network elements come from different manufacturers, exacerbates the complexity of mobile communication networks.

Using the three domestic operators (China Mobile, China Telecom, and China Unicom) as examples, the current network possesses millions of base stations and hundreds of thousands of network equipment, and the total number of employees exceeds one million. Obviously, this development model of stacking generations to increase the complexity of networks will be unsustainable sooner or later.

There are two ways to solve this problem: one is to withdraw the old network equipment, which some operators have achieved in terms of 2G withdrawal; the other is to use various new technologies to simplify the implementation and deployment of the network. The former path has a longer cycle and can only solve the issue of the coexistence of multi-generation networks. The latter path is used in the design of the 5G network architecture. It can solve the problem of its own complexity and can expand some technologies to support the old system, so as to simplify the operation and deployment of the mobile communication network.

2.2.2 Virtualization/Cloudification

On October 23, 2012, AT&T and 12 other operators announced plans to start the standardization of Network Functions Virtualization (NFV) at the first SDN & OpenFlow world congress. Later, an NFV Industry Specification Group (ISG) was established within the organizational framework of the European Telecommunications Standards Institute (ETSI). The purpose of NFV is to migrate a variety of differentiated special hardware network devices to general and standardized high-performance servers, switches, and storage devices with the help of virtualization technology in the IT industry, so as to achieve the decoupling of network functions and special hardware. Network functions can be flexibly deployed on the general hardware platform and can be instantiated or migrated on demand with the network location, as well as dynamically and elastically expanding or shrinking the capacity to enable the rapid launch of new network features. The NFV technical framework [3] defined by the ETSI NFV ISG is shown in figure 2-4.

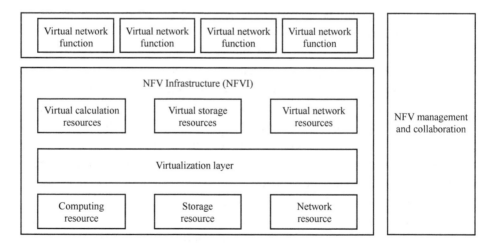

Figure 2-4 The NFV technical framework defined by the ETSI NFV ISG

The original intention of virtualization was to replace dedicated hardware with general-purpose hardware, and the starting point was highly desirable. While mobile communication networks have demanded performance, cost, and energy requirements, after several years of practice, NFV has succeeded in introducing more vendors (servers, storage, and virtualization software), greatly increasing the difficulty of system integration and challenging the skills of O&M staff. Given the results of some operators' commercial practice, it did not reduce the Capital Expenditure (CAPEX), and the Operational Expenditure (OPEX) has increased, which at least proves that the full adoption of common hardware is not economical in the short term.

Therefore, industry has put forward a more pragmatic cloud concept based on virtualization. There is no particularly unified definition of the cloud. Generally speaking, it is reflected in the pooling of hardware resources, the full distribution of software, and the full automation of operation and maintenance management. Compared with virtualization, the starting point of the cloud is resource pooling. Unlike virtualization, which must be based on general hardware, the cloud is more in line with the actual needs and technical development trend of current mobile communication networks and has become the industry consensus.

2.2.3 Software-defined network

Software-defined network technology is represented by the OpenFlow protocol [4] which is defined by the Open Networking Foundation (ONF) and related technologies. Through the OpenFlow protocol, the functions of the traditional router and switch are divided in two. The control plane functions are concentrated in the OpenFlow controller, and the user surface functions are concentrated in the OpenFlow switch. The interface protocol between the two is OpenFlow. The specific technical framework is shown in figure 2-5.

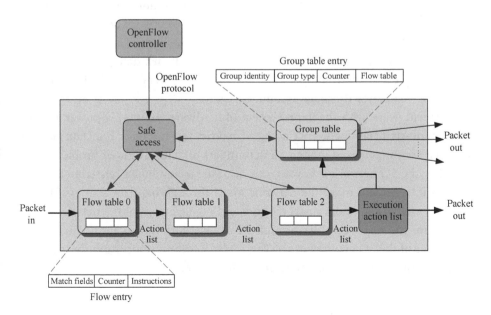

Figure 2-5 The technical framework of OpenFlow

The OpenFlow controller controls the behavior of the OpenFlow switch through the Flow Table. The Flow Table is stored in the switch and consists of several Flow Entries. Each flow entry usually includes Match Fields and Instructions. To be more specific, the matching field usually includes the information to be matched (which can be various message header fields or the port number of the switch receiving the message) and the specified value of the information to be matched. The specified value is used as a comparison with the actual value of the message's information to be matched, and the instruction is used to indicate the action to operate the message.

Through the cooperation of the OpenFlow controller and OpenFlow switch, the functions of traditional network devices such as routers and switches can be easily achieved. Meanwhile, the functions of the control plane are concentrated in the controller, which is convenient for macro control of the network. Moreover, user interface functions are simpler and can be deployed flexibly on demand.

Although OpenFlow itself has not been a commercial success, its impact has been far-reaching. Major equipment vendors have introduced SDN fixed network equipment with the idea of control-and-forward separation, bringing a technological revolution to the fixed network market. Mobile networks introduced CUPS features in 4.5G, also drawing on SDN ideas, and 5G network architectures can inherit and continue this part of the design.

2.2.4 Intellectualization

The mobile communication network contains tens of millions of network devices, providing services for billions of users. These massive network devices constantly generate a large amount of data, including logs, alarms, and performance monitoring data. How to make full use of these network data has always been an issue of great concern to operators. Obviously, it is very unrealistic to rely on manpower to analyze this huge amount of data.

In recent years, with the improvement of hardware computing power and the progress of Artificial Intelligence along with other technologies, Big Data analysis has been commercialized in increasing numbers of fields. A major problem to be solved by the 5G network is how to introduce the technologies related to big data analysis into 5G and make the network smart through big data analysis, that is, to achieve the automation, smart control, and management of the network, so as to improve the resource utilization of the 5G network and improve the service experience of the 5G network users. This is also a technical direction for the long-term evolution of network architecture in the future.

2.2.5 Multi-access

With the development of network technology, there are more types of access technologies. In addition to the access technologies defined by 3GPP, there is also a variety of non-3GPP-defined access technologies, such as Wi-Fi, and fixed broadband access, including Digital Subscriber Line Access Multiplexer (DSLAM), the Ethernet, optical fiber, and cable TV access. For integrated operators, the most reasonable choice is to make full use of the advantages of different access technologies to meet different user needs.

The coexistence of multiple access modes meets the needs of a variety of users and also makes the network increasingly complex. Since different access methods use specific network architectures and protocols, different network devices are defined, especially when the vast majority of traditional network devices still use special hardware. This leads to a large number of different hardware "boxes" in the network, which poses great difficulties for network operation and the maintenance of operators, resulting in high OPEX.

Clearly, if multiple access technologies can be supported simultaneously through a single network architecture, thereby replacing the network-side equipment of other access technologies, the types of equipment in the network can be greatly simplified, thus simplifying the deployment and operation of the network.

2.2.6 Conclusion

The mobile network has become very large and complex since its commercialization from 2G to 4G, and operators are eager to simplify mobile network operations and deployments to reduce OPEX while improving user experience. It has taken almost a decade for 4G network architecture to evolve into 5G. As new technologies have emerged during this decade, what follows is a brief introduction to the industry's consensus on the technologies that will have a significant impact on simplifying mobile network operations and deployment. These technological directions have also had a profound impact on the standards definition process for the 5G network architecture. It is worth noting that many of the new technologies were introduced in the CT field only after their application in the IT field had matured. This reflects the fact that the CT field is very open from the technical point of view, constantly drawing on the successful elements of other fields. It is also because the CT field carries more of an infrastructure function and is very cautious about the selection and adoption of new technologies.

2.3 Design principles of the 5G network architecture

2.3.1 Introduction

In the face of diversified service needs and the expectations of operators who want easier deployment and operation, 5G network architecture has introduced a large number of new technologies in its design, driving the mobile network architecture to develop in a smart and flexible direction. In the 2G/3G/4G era, a network architecture was defined to support all applications, while the 5G era is more about defining a basic architectural framework on top of which the required network functions can be customized and overlaid on demand according to the needs of different applications and industries. Through the joint efforts of the industry's wireless network architecture experts, 3GPP finally determined the design principles of 5G network architecture such as functional modularity, interface service-oriented, control and forwarding separation, access-independent, and flexible anchor points [5], as shown in figure 2-6.

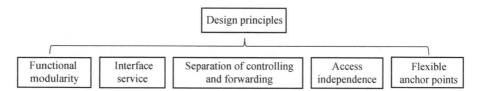

Figure 2-6 The design principles of the 5G network architecture

2.3.2 Function modularity

The 5G architecture needs to support three application scenarios: eMBB, mIoT, and URLLC, and must meet the differentiated needs of operators and various vertical industries. Therefore, the design of the network architecture must be flexible enough to be customized according to specific needs. In this way, modular design has become the most reasonable choice. By decoupling the complex network functions and decomposing them into multiple independent modules according to their functions, each functional module can be implemented in a variety of simplified or complex ways. Ideally, it can be assembled according to the actual networking requirements, like building blocks.

The idea of modularity is easy to understand, but the difficulty lies in the fact that different functions of mobile communication networks are often coupled in some. For example, there is an inextricable link between mobility management and session management, which is difficult to decouple. In addition, the granularity of decoupling is also critical. The larger the module, the smaller the gain of flexibility brought by decoupling design, while the smaller the module, the more types of modules there are, which in turn will increase the complexity of assembly. Therefore, a balance between different decoupling schemes is needed to make the network architecture flexible enough without being too complex.

2.3.3 Interface service

There are numerous interfaces defined in the 4G network architecture, and different protocols are used for different interfaces. A network element usually needs to interact with multiple other network elements, resulting in a network element usually having to support multiple protocols. Moreover, these different interface protocols are bound together through the signaling process, and it is difficult to directly reuse signaling for the interaction between network element A and network element B to the interaction between network element A and network element C, which are usually defined separately and independently.

The 5G network architecture expects that on the basis of the modularity of network functions, the interaction interface of each network function with other network functions will be defined as a service, so that it can be invoked by all other network functions that need it. For example, for network function A, just define the services A1, A2, and A3 that A provides externally, so that network function B can invoke services A1 and A2, and network function C can invoke services A2 and A3, without having to define them separately for different interaction objects.

The use of service-enabled interfaces in the 5G network architecture is a huge change. In order to ensure the timely completion of the 5G standard development, 3GPP finally decided to restrict the principle of service-enabled interfaces to the interaction between Core Network elements. The interaction between Core Network elements and the Access Network still uses traditional signaling interfaces, so that except for the Core Network elements that interface with the Access Network, other Core Network elements only need to support a service-enabled interface without supporting other protocols.

The interface service solves the problem of inconsistent interface protocols in the original network architecture. Meanwhile, the original mode of defining interactive signaling according to the interface is changed to defining services according to the network functions, which significantly reduces the workload of standard definition, development, and maintenance.

2.3.4 Separation of controlling and forwarding

In mobile communication networks, the service models of the control plane and forwarding plane are very different. Decoupling the control plane and user plane can bring many benefits:

(1) The control plane and forwarding plane can be upgraded, expanded, and reduced independently.
(2) The control plane can be deployed and maintained centrally.
(3) After simplifying the forwarding plane, it is more suitable for distributed deployment and can optimize routing and shorten the path of the forwarding plane to improve user experience.

The soft switch system of the CS network takes the lead in realizing the decoupling of control and forwarding. Its commercial success proves its rationality. The forwarding plane data traffic of the PS network is much larger than that of the CS network, and the services carried by the PS network are far richer and more diverse than that of the CS network. It is bound to require the forwarding plane to provide greater throughput and the control plane to provide richer control functions. Against this background, the decoupling of control and forwarding in the PS network is more necessary.

In fact, the decoupling of control and forwarding has been an important direction in the evolution of 3GPP PS network architecture. The MME elements in 4G networks are pure control-plane elements, while S-GW/P-GW are more user-plane oriented but still integrate a lot of control functions. 4.5G architecture introduces the CUPS feature, which further decomposes S-GW/P-GW into gateway control-plane GW-C and gateway user-plane GW-U, and it has been successfully commercialized in some operators' networks, while 5G architecture will continue to inherit this basic design principle.

2.3.5 Access independence

As networks evolve, the coexistence of multiple Access Networks leads to increasingly complex mobile communication networks. 3GPP has been committed to incorporating different access technologies into the scope of 3GPP standards to achieve a unified network architecture, from the early Interworking WLAN (I-WLAN) to Universal Mobile Access (UMA) and Generic Access Network (GAN) technologies. Access (UMA) and Generic Access Network (GAN) technologies, 3GPP has continued to carry out a lot of research and standards development work. Since the Rel-8 EPS protocol was developed, non-3GPP access has been one of the key research areas of the standard. 3GPP TS 23.402 [6] protocol has established two basic frameworks for trusted access and untrusted access for non-3GPP access technologies such as Wi-Fi, laying the foundation for fixed-mobile network convergence.

To achieve true fixed-mobile convergence, the network architecture needs to be designed to be access independent and compatible with LTE, NR, Wi-Fi, and various fixed broadband access technologies such as fiber, DSLAM and Cable. This places high demands on access authentication, session management, QoS model and policy control of the network architecture. Access independence is ultimately reflected in the use of unified interface signaling between the terminals of different access technologies, Access Networks and Core Networks. In different access technologies, there may be differences in the details, but the overall protocol mechanism is only one set.

2.3.6 Flexible anchor points

One of the advantages of controlling, forwarding, and separating is that the user interface becomes simple, so it can be deployed flexibly according to actual needs. Increasingly, mobile Internet applications require the shortest possible delay, while the network delay is directly proportional to the transmission distance of the network link. In order to reduce the delay, it is necessary to deploy the user surface anchor near the user location. However, if the deployment of user face anchors is too low, it will increase the construction cost of the entire network and will also make it difficult to ensure service continuity when users move in a large range. Therefore, some user face anchors need to be deployed in higher network locations.

In addition, in roaming scenarios, in order to access the services of the home network, the user-plane anchor of the home network needs to be selected, while in order to access the services of the visiting network, the user-plane anchor of the visiting network needs to be selected. 5G network architectures need to be designed to be able to support all these scenarios.

2.3.7 Conclusion

The design principles of 5G network architecture must fully consider the application scenarios and positioning of 5G networks and must also focus on the urgent expectations of operators to simplify network operation and deployment, while incorporating the latest achievements in ICT technology. In addition to the above basic principles, there are also some auxiliary design principles, such as support for stateless network elements, thus decoupling computing and storage capabilities, and support for bi-directional capability opening, enabling better collaboration between networks and applications.

References

1. ITU. ITU-R M.2083-0 Recommendation, IMT Vision—Framework and overall objectives for the future development of IMT to 2020 and beyond [Internet]. 2015

Sept [cited 2019 Dec 8]. Available from: https://www.itu.int/dms_pubrec/itu-r/rec/m/ R-REC-M.2083-0-201509-I!!PDF-C.pdf

2. Huawei. White Paper on Top 10 Application Scenarios in the 5G Era [Internet]. 2017 Dec [cited 2019 Dec 8]. Available from: https://www.huawei. com/cn/industry-insights/outlook/mobile-broadband/insights-reports/5G-unlocks-a-world-of-opportunities.

3. ETSI. Network Functions Virtualization (NFV); Architectural Framework [Internet]. 2014 Dec [cited 2019 Dec 8]. Available from: https://www.etsi.org/deliver/etsi_gs/ NFV/001_099/002/01.02.01_60/gs_ NFV002v010201p.pdf

4. ONF. OpenFlow Switch Specification [Internet]. 2012 Jun [cited 2019 Dec 8]. Available from: https://www. opennetworking.org/wp-content/uploads/ 2014/10/ openflow-spec-v1.3.0.pdf

5. 3GPP. Technical Report 23.799: Study on Architecture for Next Generation System [Internet]. 2016 Dec [cited 2019 Dec 8]. Available from: http://www.3gpp.org/ftp/ Specs/archive/23_series/23.799/

6. 3GPP. Technical Specification 23.402: Architecture enhancements for non-3GPP accesses [Internet]. 2019 Jun [cited 2019 Dec 8]. Available from: http://www.3gpp. org/ftp/Specs/archive/23_series/23.402/

CHAPTER 3

Key Features of 5G Network Architecture

3.1 Overview of 5G network architecture

Based on the 5G network structuring principles introduced earlier, 3GPP SA WG2 defines 5G network architecture and related service processes [1–3].

5G network architecture is designed based on the separation of control and forwarding. Its structure can be divided into a control plane and a user plane (or data plane) according to the basic functions of the Network Element (NE). The user plane is responsible for forwarding and processing user packets, including the forwarding function of the base station and one or more user plane functions (UPF). The control plane is in charge of the access authentication, mobility management, session management, and policy control of the UE.

The control plane functions of the 5G network architecture are designed based on the principle of servitization. Each Core Network's NEs from the control plane provides the public with a servitized interface based on HTTP protocol. Control plane NEs communicate with each other by invoking each other's servitized interface. These servitized relationships are combined through standardized order and parameters to form the various service control processes of the 5G network. See section 3.3 of this book for specific design principles of servitization architecture.

The main NEs and functions of the 5G system are as follows:

(1) Application Function (AF): to interact with other control network elements of the 5G network on behalf of Application, including providing service QoS strategic requirements, and routing strategic requirements.

(2) Access and Mobility Management Function (AMF): responsible for User Access and Mobility Management. The functions of AMF are complex, including ending non-access-stratum (NAS) signaling security, user registration, accessibility, mobility management, N1/N2 interface signaling transmission, access authentication, and authorization.

(3) Access Network (AN): the 5G Access Network includes Next Generation Radio Access Network (NG-RAN) or non-3GPP access connected to the 5G Core Network.

(4) Authentication Server Function (AUSF): for the authentication of users' access to 5G network.

(5) Data Network (DN): a specific data service network accessed by UE. In 5G networks, DN is identified by a Data Network Name (DNN). A typical DN includes the Internet and IMS.

(6) Network Exposure Function (NEF): provides open access to the capabilities and events of the 5G network externally, as well as receiving relevant external information.

(7) Network Repository Function (NRF): provides registration and discovery capabilities for NEs in 5G networks.

(8) Network Slice Selection Function (NSSF): the ability to select network slices.

(9) Policy Control Function (PCF): responsible for generating the UE access policy and QoS Flow control policy.

(10) Session Management Function (SMF): provides UE session management (such as session establishment, modification, and release), IP address allocation and management, and UPF selection and control.

(11) Unified Data Management (UDM): for user contract management, access authorization, and the generation of authentication information.

(12) Unified Data Repository (UDR): provides the storage capability of subscribed data, policy data, and openness-related data.

(13) Unstructured Data Storage Function (UDSF): supports the storage of unstructured data of various network elements.

(14) UE: the terminal equipment used by the user.

(15) User Plane Function (UPF): provides user plane functions such as user message forwarding, processing, connection with DN, session anchor, and QoS policy enforcement.

Figure 3-1 shows the servitized 5G network architecture in non-roaming scenarios.

In the above servitized architecture, each Core Network control plane NE provides a servitization interface named after it externally. For example, AMF provides Namf interface externally. According to the principle, each NE's servitized

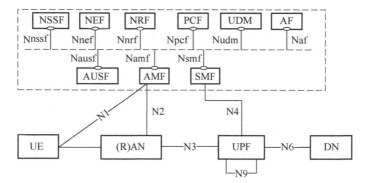

Figure 3-1 5G network architecture in non-roaming scenarios (servitized interfaces)

Note: Some service-specific NEs, such as UDR, UDSF, Network Data Analytics Function (NWDAF), and Changing Function (CHF) are not shown in the figure above.

interface can be invoked by any external NE. However, in 5G architecture, due to the limitations of standardized processes, protocol design, and security, the calling and called relationship of each specific servitized interface will be crystallized into a clear point-to-point connection relationship. For example, the servitized interface of SMF will be called by AMF and AF but will not be called by unrelated NEs on processes such as AUSF according to the process design of Rel-15/16. In order to reflect the nature of servitized interfaces and describe the constraints on the invocation of these interfaces, the 5G network architecture defines the representation of both servitization and traditional point-to-point communication.

It is important to note that the service-oriented architecture diagram and the point-to-point architecture diagram are just different representations of the same 5G network architecture. The servitized architecture diagram reflects the ability of the 5G Core Network control plane network elements to provide services to the outside world, while the point-to-point architecture diagram reflects the standardized invocation relationships between network elements in the current version of the defined standardized process. Similarly, the serviced interface names in the point-to-point architecture diagram are the result of the specific instantiation of the serviced interface names in the serviced architecture diagram in the current version of the standardization process. For example, the N11 interface between AMF and SMF is concretely implemented through the services of the two serviced interfaces Namf and Nsmf. The 5G system architecture for a non-roaming scenario in the form of a point-to-point interface is shown in figure 3-2.

Compared with 4G, a major change in the function of 5G Core Network elements is the separation design of mobility management and session management. The mobility management and session management functions of MME and Serving GW/PDN GW (or their control plane) in 4G networks are refactored and assigned in 5G network architecture and are implemented by AMF and SMF respectively. The main driving forces behind the separation of mobility management and session management in 5G networks include

(1) by taking session management out of the basic flow of access control in 5G networks, certain IoT-like terminals can communicate through the control plane after accessing the network, instead of maintaining a session all the time. The separation of AMF and SMF helps achieve independent management and maintenance of mobility and session managements;

(2) within the network slice architecture, terminals may access multiple slices at the same time. In this case, mobility management for multiple network slices is still performed at the UE granularity, but session management can be performed based on the slice granularity. This leads to different granularity of slicing deployment for mobility management network elements and session management network elements, which can be found in section 3.4 of this book. Separating these two functions is designed to facilitate more flexible network deployments; and

(3) in different deployment scenarios, the deployment locations of AMF and SMF may be different, and the separation of functions is more conducive to the deployment of these functions in different locations as required.

In Rel-15/16 standard protocol, UE, AN, and UPF do not support a servitized interface. The AMF and SMF serve as the interfaces between the control plane and UE, AN, and UPF respectively, and provide traditional point-to-point interfaces N1, N2, and N4 externally.

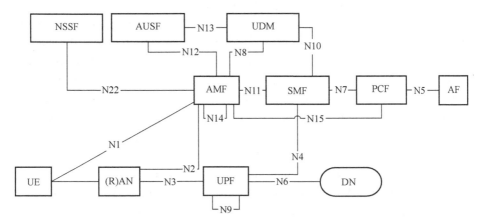

Figure 3-2 5G system architecture in non-roaming scenarios (point-to-point interfaces)

When the UE roams from the attributed PLMN to the visiting PLMN, the roaming scenario can be divided into roaming Local Breakout scenario and roaming Home Routed scenario according to whether the user side of the UE is visiting PLMN terminated or attributed PLMN terminated. In the roaming scenario, the visiting PLMN (VPLMN) and the home PLMN (HPLMN) interoperate with each other through their respective Security Edge Protection Proxy (SEPP) for service-based interfaces to achieve message filtering and topology hiding functions across PLMN control-plane interfaces.

Figures 3-3 and 3-4 show different types of interfaces in the 5G system architecture in a Local Breakout scenario.

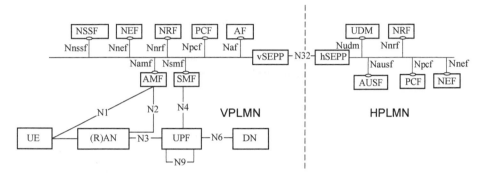

Figure 3-3 5G system architecture in a roaming Local Breakout scenario (with servitized interfaces)

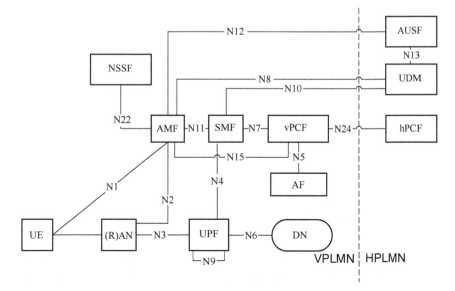

Figure 3-4 5G system architecture in the roaming Local Breakout scenario (with point-to-point interfaces)

Figures 3-5 and 3-6 show the different interface types of the 5G system architecture in a Home Routed scenario.

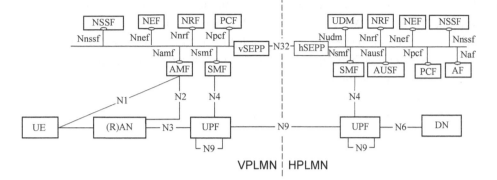

Figure 3-5 5G system architecture in the roaming Home Routed scenario (with servitized interfaces)

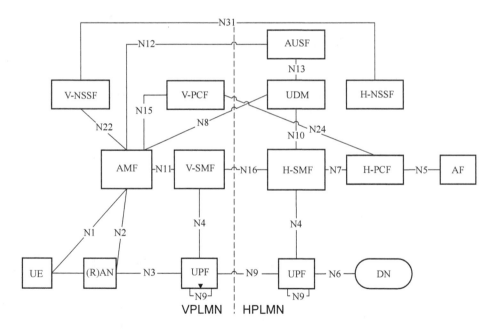

Figure 3-6 5G system architecture in the roaming Home Routed scenario (with point-to-point interfaces)

In a Cloud-based deployment environment, most of the control plane network elements can operate in a virtualized form. The 5G Core Network requires more consideration than traditional cellular network architecture on how to achieve efficient maintenance and access to the user context while allowing dynamic instantiation and de-instantiation of network elements. Therefore, a data storage architecture based on UDSF and UDR has been specifically designed for the 5G Core Network.

For example, UDSF serves as an unstructured database that can be used to hold the internal data of any network element. The interface definition of UDSF is shown in figure 3-7.

Figure 3-7 Interface definition of UDSF

The purpose of UDSF is to realize efficient information sharing between devices from the same manufacturer based on shared storage, and thus realize stateless processing of NEs. However, the complete standardization of internal information cannot be realized due to the different private implementation, internal processing logic and information storage methods of different manufacturers. Although NUDSF's data interface protocols can be standardized, the specific information conveyed in them can also be vendor-private and therefore does not need to be standardized. Depending on how the product is implemented, different NEs may share a UDSF or may use their own specific UDSF.

In addition to unstructured information, there is a lot of structured (i.e., standards-definable 3GPP) information in the network, such as subscribed data, policy data, and data obtained from AF through external open capabilities. For this type of data storage, 3GPP defines UDRs for storing structured data. UDRs are connected to UDM, PCF, and NEF through standardized interfaces. The UDR interface definition is shown in figure 3-8.

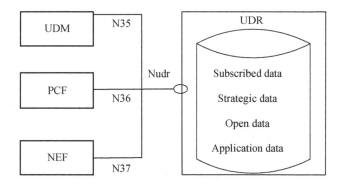

Figure 3-8 Definition of the UDR interface

Figure 3-8 shows that the UDR stores four types of data:

(1) Subscribed data
(2) Strategic data for PCF to make strategic decisions
(3) Open data for NEF to open to third parties
(4) Application data, which is used to store application-related information provided by AF to the network, such as N6 interface routing information

UDSF and UDR are both data storage devices, meaning that operators can deploy them flexibly through centralized or distributed data stores to balance maintenance complexity and storage performance.

3.2 Basic features of the 5G network architecture

3.2.1 Mobility management
3.2.1.1 Introduction
Compared with wireline networks, wireless networks can support terminal mobility more effectively. When users are connected to a wireless network (such as by using a mobile phone to access the Internet), they can move freely without being restricted by network cables and interfaces.

In order to receive or send data, UEs connected via wireless need to "listen" for frequencies. However, since UEs are often powered by a battery with limited capacity, if the UE is in "listening" mode at any given moment, the battery can be exhausted very quickly. To save power, the UE enters "idle" mode when there is no data (or other control information) to be sent or received. In this state, the UE only "listens" to a limited number of frequencies, thus saving power.

Due to the above characteristics of wireless network connection, the management of UE mobility by the wireless network is particularly important. The main content of mobility management includes managing the UE's registration and connection state, designing the transition process between different connections and registration states, and tracking the position of the UE in different connection modes as it moves and sends its data.

The transition between UE connection modes can be achieved through the corresponding mobility management processes. Table 3-1 briefly describes the relationship between the 3GPP mobility management processes from the perspective of a passenger traveling to the airport. More details can be found in section 3.2.1.2.3 of this book (Transitions between States).

Table 3-1 3GPP mobility management process corresponding to user behaviors

User behavior	Corresponding process	Description
Wakes up early in the morning and turns on cellphone	Initial registration process	A UE can use network services only after registering to the network. The UE's location information is saved on the network side. See section 3.2.1.5.1 of this book for details.
Reads the news on the bus to the airport	Switching process	When a UE in the connected state is on the move, the base station decides whether to switch to a cell or base station with a stronger signal based on the reported information from the UE. See section 3.2.1.5.2 of this book for details.
Sleeps on the bus and stops background apps	Access Network release process	As the UE does not need to continue services, it initiates the Access Network release process and goes into an idle state. See section 3.2.1.5.4 of this book for details.
Sleeps until arrival at the airport	Mobility registration process	When a UE in an idle-state moves out of the configured registration area, the mobility registration process needs to be initiated. See section 3.2.1.5.1 for details.
Sends a WeChat safety-check message to family	Service request process	When a UE in an idle-state needs to send data, it initiates the service request process to activate the corresponding data transfer channel and then sends data. See section 3.2.1.5.3 for details.
Turns off cellphone before takeoff	Deregistration process	The UE is logged out of the network due to the cellphone being turned off. See section 3.2.1.5.5 for details.

3.2.1.2 Registration and connection status management

3.2.1.2.1 Registration status

There are two registration states for UE, namely RM-REGISTERED and RM-DE-REGISTERED.

Before the UE can receive service data (e.g., when the phone is turned on), it needs to complete the registration to the network first. After the UE registration is completed, the UE is in the registered state and this registration state is saved at the UE and the AMF. For the registered UE, the AMF also needs to save the mobility management context of the UE.

When a UE is not registered in the network, the UE is in a deregistered state, and cannot communicate with normal services through the network. For a UE that has been registered in the network, even if it is in the deregistration state, the AMF can still store part of the context information of the UE's mobility management in order to simplify the process of subsequent UE registration.

For UEs with both 3GPP access and non-3GPP access, the registration status management corresponding to these two accesses is independent of each other.

3.2.1.2.2 Connection status

After the UE is registered to the network, further Connection Management (CM) is required for it, including the establishment and release of the NAS signaling connection between the UE and AMF. The NAS signaling connection consists of two parts: the connection between the UE and the Access Network (AN), and the N2 connection between the AN of the UE and the AMF. The NAS signaling connection between UE and AMF has two states, namely CM-IDLE and CM-CONNECTED. Similar to the management of registration status, the management of connection status is independent for 3GPP and non-3GPP access.

When the UE is in the idle state, the connection between the UE and the AN, the N2 connection between the AN and the AMF, and the N3 connection between the AN and the UPF are released. Figure 3-9 is a diagram of the N2/N3 and 5G-AN connection states of the UE in the CM-IDLE state.

The UE initiates a transition from the idle state to the connected state when sending the initial NAS message. The UE is considered to enter the connected state when a signaling connection is established between the UE and the AN. For the AMF, the AMF considers the UE as entering the connected state when an N2 connection is established between the AN and the AMF. Figure 3-10 depicts the schematic diagram of the N2 and 5G-AN connection states of the UE in the CM-CONNECTED state.

The transition of the UE from the idle state to the connected state needs to be achieved through processes such as service request process or registration process. In order to release radio resources on demand and avoid complex processes, a new sub-state of the connected state, called Connection Management Connected

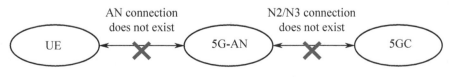

Figure 3-9 Schematic diagram of the CM-CONNECTED state

Figure 3-10 Schematic diagram of the CM-CONNECTED state

Radio Resource Control Inactive (CM-CONNECTED RRC Inactive), or RRC Inactive state for short, has been introduced in 5G systems for UEs with 3GPP access.

The NG RAN releases the air interface connection resources of the UE in the RRC Inactive state. Further, the AMF does not sense whether the UE is in the RRC Inactive state unless the AMF subscribes to the NG-RAN node with a notification message of the UE's conversion to RRC Inactive. In the RRC Inactive state, the NG-RAN node that last served the UE maintains the context, N2 connection, and N3 connection corresponding to that UE. Figure 3-11 is a diagram of the N2 and RRC connection states of the UE in the RRC Inactive state.

Figure 3-11 Schematic diagram of the RRC Inactive state

The AMF provides RRC Inactive Assistance Information (RIAI) to the NG-RAN node to help the RAN node determine whether the UE can change from a connected state to RRC Inactive state. RIAI includes the registration and update of the timer and the UE identity index values for the registration area and period configured for the UE. It also includes the UE's specific discontinuous receive period, and whether Mobile Initiated Connection Only (MICO) is activated. It is important to note that even if the AMF provides this information, the NG-RAN node can decide based on local policy not to put the UE into the RRC Inactive state.

When the NG-RAN node determines that it needs to change the UE to the RRC Inactive state, the NG-RAN node needs to notify the UE of the RRC Inactive state. Table 3-2 shows the RRC Inactive information that the NG-RAN node provides to the UE.

Table 3-2 RRC Inactive information provided to the UE by the NG-RAN node

Information	Description
RAN Notification Area (RNA)	The NG-RAN node determines what registration region the RNA should contain based on the UE registration region. For details, see section 3.2.1.3.1.
Paging Occasion	When setting the paging period, the NG-RAN node will be generated by combining the UE-specific discontinuous receiving period and the UE identification index value.
RAN Notification Area Update (RNAU) Timer	The RAN node determines the value of the periodic RNA update timer according to the periodic registration update timer, and the corresponding time length of the periodic registration update timer is longer than the value of the periodic RNA update timer.
Index Value of the UE Context on the RAN Side	The UE provides this information to the RAN the next time the RRC Inactive connection is restored, for the RAN to locate the context to the UE.

3.2.1.2.3 Transitions between States

The transition between different states and the mobility management process is shown in figure 3-12, which depicts the transition between different states of the UE and the associated mobility management process. The UE in the deregistered state is registered to the selected PLMN through the Initial Registration process before receiving the service, and in the registered state, the UE updates the location information of the UE through the Mobility Registration Update, the Periodic UE notifies the network that the UE is still active through Mobility Registration Update. When the UE no longer needs to be registered to the network, the UE or AMF performs a Deregistration process. In addition, the AMF may perform an implicit Deregistration when the AMF considers that the UE is no longer active (e.g., the UE has not contacted the network after a certain preset time). After performing the deregistration process, the UE enters the deregistration state.

When the UE has registered with the network and is in an idle state, if the AMF has downlink signaling or data to send to the UE, the AMF will initiate the Service Request process triggered by the network and then trigger the paging process to the UE. If the UE receives a network pager or has uplink signaling or data to send, it must perform the Service Request process to enter the connection state. The UE in the connected state enters the idle state when the AN signaling connection is released. In addition, when the UE needs to change the connected base station due to movement, a Handover process will be initiated.

See section 3.2.1.5 of this book for the above mobility management processes.

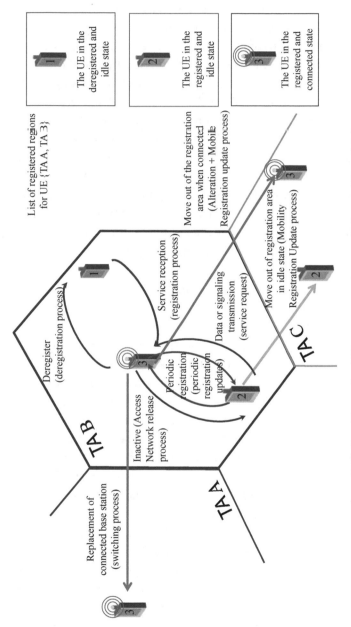

Figure 3-12 Transition and mobility management process between different states

3.2.1.3 Reachability management

As mentioned earlier, the introduction of the idle state can save the power consumption of the terminal, but it can also lead to the network not being informed of the exact location of the UE. When the UE is in an idle state, the location tracking of the UE relies on the network's reachability management of the UE. Reachability management is responsible for detecting whether the UE is reachable and providing the network with the location of the UE. Reachability management is achieved by paging the UE and location tracking of the UE.

The UE and the AMF negotiate the relatability characteristics of the corresponding UE during the registration process. For the UE in the idle state, there are two types of relatabilities. One is allowing sending downlink data or signaling to a UE in an idle state. For that type of UE, the network can know the UE's location from the granularity of the Tracking Area (TA) and the AMF can page the UE. The other is the MICO mode, when sending downlink data and signaling is not allowed for a UE in an idle state. When the UE is in MICO mode, the network can only send downlink data to the UE when the UE is in the connected state. When the UE is in an idle state, the network cannot page the UE.

3.2.1.3.1 Reachability management for the UE

Relatability needs to be managed on the temporal and spatial dimensions. From the spatial point of view, the network manages the activity range of a UE in the idle state using registration zone management. Within this range of activity, the UE can move freely without interacting with the network due to reachability management. From the temporal point of view, the network manages the UE in an idle state using periodical timer management. The UE must inform the network about its reachability when the corresponding timer expires.

For a UE in the connected state, the AMF stores the information of the AN node, providing the connection for the UE. In addition, if the AMF subscribes to the NG-RAN with the location information of the UE at cell granularity (see section 4.10 of 3GPP TS 23.502 [2] for details), the AMF can obtain information on the specific cell where the UE resides.

In order to manage the range of areas where the UE can be discovered in the registration state, the 3GPP TS 23.501 standard defines the concept of Registration Area (RA), where the AMF provides the UE with the registration area information during the registration process, which consists of a complete TA. When the UE is in an idle state, the AMF senses the location of the UE at the granularity of the registration area. When there is downlink data or signaling to be sent to the idle UE, the AMF will page for the UE within the registration area. When the UE moves out of the registration area, the UE initiates a mobility registration process.

During this process, the AMF assigns a new registration area to the UE based on the latest location of the UE.

During the registration process, the UE receives from the AMF the value for the periodic registration update timer. When the registered UE enters an idle state, the UE starts the periodic registration timer. After the periodic registration timer times out, the UE performs a periodic registration to indicate to the AMF that it is still reachable. If the UE is out of network coverage at the time of the periodic registration timer timeout, the registration process needs to be performed when the UE next returns to coverage. On the AMF side, when the UE in the registered state becomes idle, the AMF runs a mobile reachable timer for the UE with a duration slightly larger than the value of the periodic registration timer assigned to the UE.

If the connection state of the UE in the AMF turns to the connected state, the AMF will stop the mobile reachable timer. If the timer expires, the AMF deems the UE to be unreachable. However, because AMD does not know whether the UE is just temporarily unreachable (for example, temporarily moving to somewhere with bad network coverage) when the mobile reachable timer expires, it would not immediately deregister the UE. After the mobile reachable timer expires, the AMF will clear the Paging Proceed Flag (PPF) and initiate the implicit detach timer. If the UE turns back to the connected state before the timer expires, the AMD should stop the implicit detach timer and set the PPF. Otherwise, when the timer expires, the AMF initiates the implicit detach process of the UE. The state change of the AMF and the UE when the timers expire are shown in figure 3-13.

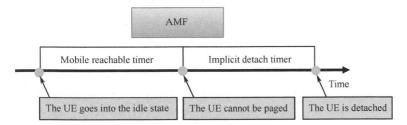

Figure 3-13 The state change of the AMF and the UE when the timers expire

There is a strong relationship between the connection state of the UE and the UE's reachability. For example, for a connected UE, the network can be informed of its location exactly. However, when the UE moves, its switching process or the location reporting process needs to be triggered, which leads to additional signaling overhead. In general, the maintenance cost of the UE increases with the granularity of reachability management, so it is important to consider the balance between them.

When the UE mode registers to the network in MICO, the network only knows of the registration of the UE but not its accurate location. The UE only needs to initiate periodic registration after the periodic registration timer expires. For the UE, it is only necessary to initiate the cycle registration after the cycle registration timer expires. For a UE registered to the network and not in MICO mode, the network knows the location range of the UE at the RA level when the UE is in the idle state. At this point, the UE and the network need to handle mobility registrations triggered due to the UE moving out of the RA and periodic registrations triggered after the expiration of the periodic registration timer. When the UE is in a connected state, the network knows the NG-RAN node to which the UE is connected, and when the UE needs to change the connected NG-RAN node, the UE and the network need to handle the associated switching process and mobility registration. Further, when the network subscribes the cell location of the UE to the NG-RAN node, and the UE changes the cell it resides in, in addition to the switching process and mobility registration, the NG-RAN node may also report the latest cell information of the UE's residence via N2 messages. The UE reachability and the state maintenance process involved are shown in table 3-3.

Table 3-3 The reachability of the UE and the involved state maintenance process

UE reachability level	Process involved in UE state maintenance
CN level (MICO)	Periodic registration
RA level	Periodic registration/mobility registration
RAN node level	Switching process/mobility registration
Cell level	Position report/switching process/mobility registration

As table 3-3 shows, the coarser the granularity of the reachability level, the less frequently the process of maintaining the UE state is triggered. However, accordingly, the network only knows the location information of the UE at a coarse granularity, so the paging range for the UE becomes larger and more costly.

In addition, the UE state maintenance cost is also related to the UE mobility speed; the higher the UE mobility speed, the higher the state maintenance cost in terms of location update signaling. For example, a stationary UE (e.g., equipment in a server room) does not need to involve mobility registration triggered due to movement, whereas UEs that move faster (e.g., passengers in vehicles or high-speed trains) will have relatively more frequent switching processes or mobility registrations triggered.

The service traffic model also affects the signaling cost of paging, i.e., the more frequent the downlink data or signaling triggers in idle mode, the more paging signaling costs may be required.

Therefore, when determining the UE state and the UE reachability, many factors such as UE capabilities, mobility patterns, and call models need to be considered in order to find a signaling load balancing point. This is one of the main requirement scenarios for on-demand mobility management that will be mentioned in section 3.2.1.4 of this book.

3.2.1.3.2 Reachability management of the RRC Inactive state

Similar to the idle state, in order to manage UEs in the RRC Inactive state, the standard defines an RNA for NG-RAN management. The RNA can be composed of cells, TAs or RAN nodes. The NG-RAN senses the location of UEs in the RRC Inactive state at the granularity of the RNA. When downlink data or signaling is sent to the NG-RAN, the NG-RAN pings within the range of the RNA. When a UE in the RRC Inactive state enters a cell that does not belong to the RNA of that UE, the UE will perform an update of the RNA (for more details, see section 9.2.2.4 of 3GPP TS 38.300 [4]) so that the NG-RAN can assign a new RNA to the UE based on information such as the latest location of the UE.

If the UE is in RRC Inactive state and its last resident NG-RAN receives UE downlink data from the UPF or UE downlink signaling from the AMF, this NG-RAN will page the UE in the cell corresponding to the RNA. If the RNA contains cells of an adjacent NG-RAN, the NG-RAN node may send messages for paging to the adjacent NG-RAN node. Reachability management for the RRC Inactive state is shown in figure 3-14.

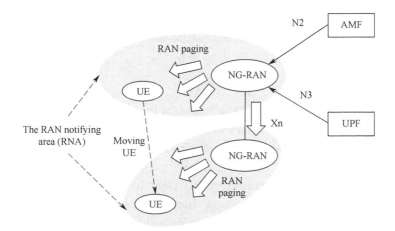

Figure 3-14 Reachability management of the RRC Inactive state

When the UE is in the RRC Inactive state, the NG-RAN node manages the reachability of the UE. When the NG-RAN node instructs the UE to change to the RRC Inactive state, the NG-RAN node will allocate a periodic RNAU timer for the UE. After the UE receives the message, it goes into the RRC Inactive state and starts this timer. When the periodic RNAU expires, the UE in the RRC inactive state executes the RNA update procedures to prove its accessibility to the NG-RAN. When the RNA update timer (the duration of which is slightly longer than the periodic RNAU timer given to the UE) expires, the NG-RAN node will initiate the detachment procedures of the AN. The NG-RAN node can provide the elapsed time between the last contact of the NG-RAN node and the UE to the current time to the AMF, so that the AMF can accurately calculate the value of the mobile reachable timer used to maintain the connecting status of the UE.

If the AMF receives the UE context detachment message from the RAN about the elapsed time mentioned in the previous paragraph, the AMF deduces the value of a new mobile reachable timer based on the elapsed time received from the RAN and the value of the normal reachability timer. Figure 3-15 illustrates how the AMF updates the mobile reachable timer.

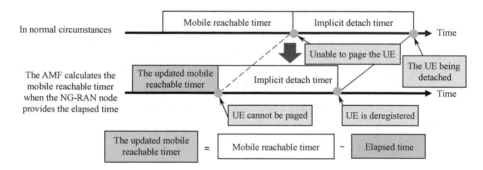

Figure 3-15 The AMF updating the mobile reachable timer

3.2.1.3.3 MICO mode
Some special UEs may have only uplink data (or their downlink data is sent immediately adjacent to the uplink data transmission, i.e., "uplink pulls downlink"). 5G networks define MICO patterns to optimize mobility management for this type of UE.

The UE can indicate that it can enter MICO mode during the initial registration or mobile registration update process. The AMF determines whether to allow the UE to enter MICO mode based on information such as local configuration, expected UE behavior, the UE indicated preferences, the UE contract information, and network policy, and indicates this to the UE in a registration accept message.

The UE and the AMF renegotiate the MICO mode during each subsequent registration process.

When the UE is in MICO mode and in the idle state, the AMF does not initiate the paging of the UE. Therefore, the registration area of the UE can be unlimited by the size of the paging area. For example, if the AMF service area is the entire PLMN, the AMF can set the registration area of the UE to "All PLMN." In this case, the mobility registration update process is not triggered as long as the UE moves within the PLMN area.

When the UE is in MICO mode and the idle state, the AMF always considers the UE as unreachable. At this point, the AMF does not trigger processes such as paging because of the need to send downlink data or signaling to the UE. Only when the UE is in the connected state will the AMF consider the UE to be reachable. A UE in MICO mode and in the idle state does not need to listen for paging. The trigger conditions for a change in the UE connection state are

(1) updating the network registration due to changes to the UE (e.g., configuration changes);
(2) the expiration of the Periodic Registration Update Timer;
(3) the unprocessed uplink data or signaling.

If the registration area of the MICO mode UE is not "All PLMN," the UE will determine whether it is in the registration area when initiating a data or signaling transmission. If it is not, the UE will first update its mobility registration.

3.2.1.4 On-demand mobility management

Different UEs may have different communication characteristics, e.g., certain monitoring devices only need to support active data uploading, and certain industry terminals can only communicate in specific areas. The network can optimize mobility management to these communication characteristics of the UE accordingly, i.e., on-demand mobility management.

On-demand mobility management consists of two aspects, namely Mobility Restriction and Mobility Pattern. With Mobility Restriction, operators can set certain areas of the network to be off-limits to a specific group of UEs. With Mobility Pattern, operators can determine the communication characteristics of UEs and set up more appropriate mobility management schemes for them, such as setting up appropriate tracking areas.

3.2.1.4.1 Mobility restriction

The main purpose of mobility restriction is to limit mobility-related processing or the service access of the UE. Its scope of action includes the UE, the NG-RAN node, and the Core Network. In general, mobility restrictions only apply to 3GPP access and wireline access, and not to other types of non-3GPP access.

When the UE is in the connected state, the Core Network provides a mobility restriction list to the Access Network. The Core Network and NG-RAN nodes perform the processing of the mobility restrictions. When the UE is in the idle or RRC Inactive state, the UE side performs Service Area Restriction and Forbidden Area processing based on the information provided by the Core Network.

When the UE is in the connected state, the Core Network provides mobility restrictions to the AN in the mobility restriction list. Mobility restrictions include RAT restriction, Forbidden Area, Service Area Restriction, and Core Network type restriction:

(1) RAT restriction: Based on the signing information, the network does not allow the UE to access a PLMN network through a restricted access technology. PLMN-based RAT restrictions need to be considered when the RAN selects the target RAT or target PLMN in the switching process.

(2) Prohibited Area: Based on the contract information, the UE is not allowed to initiate any interaction with a PLMN network in this area. The prohibited area is valid for both 3GPP and non-3GPP networks.

(3) Service Area Restriction: Based on the contract information, the Allowed Area or Non-Allowed Area of the UE when launching communications is defined. In an Allowed Area, the UE can communicate with the network normally. In a Non-Allowed Area, the UE cannot initiate service request processes or general session management-related signaling interactions with the network. When the UE is in the Non-Allowed Area, it needs to respond to paging initiated by the Core Network, NAS notification processes, and paging initiated by the NG-RAN node.

(4) Restrictions on the Core Network type: Defines the type of Core Network that allows UE access, e.g., EPC or 5GC.

The Core Network determines the mobility limits based on the UE's contract, the UE's location, and local policy information. Mobility restrictions may alter due to changes in the above information. The Allowed Area or Non-Allowed Area may also be adjusted further by the PCF based on information such as the UE location. The updated service area restrictions may be performed during the registration process or during the UE configuration update.

The network does not offer both an Allowed Area and Non-Allowed Area to the UE when offering service area restrictions; if the network offers a Non-Allowed Area to the UE, TAs in PLMNs that are not on the list are considered to belong to the Allowed Area.

In the event of a conflict between the areas defined as Service Area Restrictions and Prohibited Areas, priority will be given to the content of the definition of Prohibited Areas.

The service area restriction may contain one or more complete TAs or all TAs of the PLMN. The service area restriction is contained in the contracting data of the UE stored by the UDM and may be represented by the TA identification and/or other geographical information (e.g., latitude, longitude, or zip code). If geolocation information is used, the AMF first maps the geolocation information to the TA before sending the service area restriction information to the PCF, RAN, and UE. During the registration process, if the context of the service area restriction of the UE is not available in the AMF, the AMF obtains this information from the UDM and is able to adjust it further via the PCF. The network can update the service area restriction information via the generic UE configuration update process.

When the size of the service area restriction allocated by the AMF to the UE is limited (i.e., containing only one or more full TAs), the Allowed Area included in the service area limit provided by the AMF to the UE may be pre-configured or dynamically allocated by the AMF (e.g., dynamic TA control as the UE location changes).

When the service area restriction assigned by the AMF to the UE contains all TAs of the PLMN, the registration area assigned by the AMF to the UE will consist only of TAs belonging to the Non-Allowed Area/Allowed Area if the UE is located in the Non-Allowed Area/Allowed Area. The AMF provides the service area restriction in the form of a TA. This service area restriction may be part of the complete list stored in the UE's contract data or may be provided by the PCF to the UE during the registration process.

The AMF can limit the size of the service area restriction on the UE side by the maximum number of tracking areas allowed (this limit threshold is not sent to the UE). When the Allowed Area is used along with the maximum number, the number indicates the number of TAs that can be included in the Allowed Area. When the number is used with Non-Allowed Area, the number indicates the maximum number of TAs that can be contained in the Allowed Area outside of the Non-Allowed Area.

3.2.1.4.2 Mobile modes

The mobility pattern describes the mobility characteristics of the UE and consists of various parameters including UE capabilities, mobility speed categories, and service characteristics. The mobility pattern is a concept that AMF can use to characterize and optimize UE mobility. AMF determines and updates the mobility pattern of a UE based on one or more of the following: The UE sign-up status, the UE mobility, local network policies, and the UE auxiliary information. The statistics of UE mobility can be historical or expected UE mobility trajectories. The AMF can optimize a UE based on the mobility pattern, e.g., registration area assignment.

3.2.1.5 Key processes for mobility management

3.2.1.5.1 Registration processes

When a UE needs to access the 5G network for service data interaction, the UE initiates a registration process. The main types of registration include the following:

(1) Initial-type registration: When the UE is initially registered to the 5G system, it initiates the registration of the initial type.

(2) Mobile-type registration: The mobile-type registration process is initiated when the UE is in a connected or idle state and has moved out of the range of the original registration area, or when the UE needs to update protocol parameters or capabilities, or when the UE wants to obtain information about the Local Area Data Network (LADN).

(3) Periodic registration renewal: The UE initiates a periodic registration renewal process when the periodic registration renewal timer provided by the AMF to the UE expires.

(4) Emergency registration: Used by the UE to request emergency services.

The basic registration processes are shown in figure 3-16.

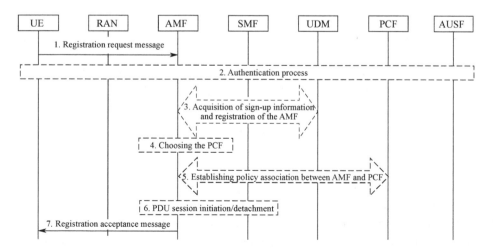

Figure 3-16 Basic registration processes

Step 1: The UE sends a registration request message to the AMF. The message
   contains the AN parameter and the NAS message of the registration request,
   and optionally the UE Policy Container information. After the RAN
   selects the AMF based on the AN parameter, it sends an N2 message to the
   corresponding AMF, which contains the NAS message of the registration
   request. See step 1 to step 7 of the registration process in TS 23.502 [2] for
   details.

Step 2: Authentication process. For the specific process, see the corresponding
   section of TS 33.501 [5]. Among these processes, the selection of the AUSF
   network element and the selection of the UDM network element are involved.
   After the authentication of the UE by the AUSF network element is passed,
   the AUSF provides the security context of the UE to the AMF.

Step 3: Acquisition of sign-up information and registration of the AMF. If the
   network selects a new AMF to provide the service for the 3GPP access of
   the UE, the new AMF needs to register it to the UDM. If the AMF does not
   have the contracted data of the UE, the AMF obtains the corresponding data
   by calling the Nudm_SDM_Get service. After obtaining the contracted data
   of the UE, the AMF establishes the UE context for the UE. See step 14 of the
   registration process in TS 23.502 [2] for details.

Step 4: The AMF selects the PCF for the UE.

Step 5: Establishing policy association between AMF and PCF, including UE
   control policy and access/mobility management policy association. Specific
   reference is made to the relevant flow for policy control and steps 16 and 21b
   of the registration flow in TS 23.502 [2].

Step 6: Protocol Data Unit (PDU) session activation/detachment process: If the
   registration request message carries the list of PDU sessions to be activated,
   the AMF will send a message to the corresponding SMF to establish the
   subscriber-plane resources. The specific process refers to the scheme of user-
   plane activation for the service request process. For PDU sessions that are
   detached locally on the UE side or not supported by the network, the AMF
   will trigger the corresponding PDU session detachment process.

Step 7: Registration acceptance message. The AMF sends a registration success
   message to the UE containing the parameters assigned to the UE by the
   network side.

The registration process described above encapsulates the steps related to initial
registration, mobility registration, periodic registration updates, and emergency
registration. It is worth noting that the authentication process described for
the second step is executed when and only when the UE performs the initial
registration, and no security context is available.

3.2.1.5.2 In-system switching process

When a terminal is connected, its surrounding environment or service may also change accordingly with the change of time and space. A 5G system may change the base station to which a terminal is connected. Specifically, the switchover may occur for the following reasons: the signal strength provided by the new base station at the location of the current terminal is somehow better than that of the original resident base station; the load on the original resident base station is higher, and load balancing is required; or the switchover is triggered by some specific service of the current terminal. The process of changing the terminal-resident base station in the connected state is called Handover.

The two switching methods involved in 5G systems are Xn switching and N2 switching. After the UE's original residing base station determines the target that needs to be switched, it will first determine whether an Xn connection exists between the two. When there is no Xn connection between the two, or when the source base station determines that the switch requires a change in the AMF providing services to the terminal, the N2 interface-based switching process will be used.

The switching process for both Xn and N2 can be divided into a preparation phase and an execution phase.

In the preparation phase, two actions are initiated:

(1) The source base station establishes a forwarding tunnel with the target base station.
(2) The target BTS establishes relevant resources for the current PDU session of the UE.

In the execution phase, two actions are also initiated:

(1) The source BTS instructs the UE to access to the target BTS.
(2) The Core Network element (or also the base station) completes the establishment of the relevant context.

Next, the Xn switching without user-plane redistribution and the N2 interface-based switching process is described separately.

The process of Xn switching without user-plane redistribution is shown in figure 3-17.

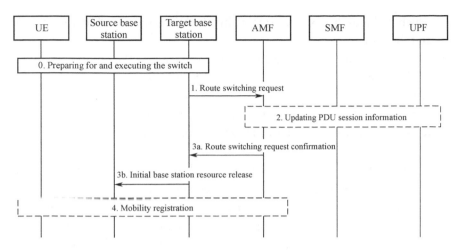

Figure 3-17 The Xn switching process without user-plane redistribution

Before the switching is performed, the source and target BTSs prepare for the switchover. The source base station needs to send information like target cell identification, access layer configuration, the UE capabilities, mapping relationship between QoS Flows and Data Radio Bearer (DRB), and PDU session context to the target base station. The target base station performs access control on the UE and determines the acceptance/rejection of certain PDU sessions based on the supported slices. Based on the received information, the target BTS establishes appropriate resources and sends a confirmation message to the source base station.

The source base station sends an RRC reconfiguration message to the terminal to instruct the terminal to make the switch. The message contains information such as the target cell identification, the new cell temporary identification, and the target base station security algorithm information. After receiving this information, the terminal performs a switching operation and sends an RRC reconfiguration completion message to the target base station.

Step 1: After the completion of the air interface switch, the target base station notifies the Core Network to switch the N3 path to the target base station. Specifically, the target base station sends a Path Switch Request message to the AMF, which contains the PDU session and its N2 information for successful switching, the PDU session and its N2 information for failed switching, and the UE location information.

Step 2: Based on the ID of the switchover success/switchover failure PDU session, the AMF finds the corresponding SMF through the locally stored UE SM context and notifies it to update the information of the related PDU session,

which includes the SMF re-establishing the N3 connection between the RAN
and the UPF.

Step 3: The AMF responds to the path switch and triggers the release of resources
on the RAN (S-RAN) side of the source base station.

Step 4: If the conditions for the registration update of the mobile type are met
(e.g., the UE moves out of the registration area), the UE initiates a mobility
registration update after the switchover is completed.

After the switchover step is executed, for the PDU sessions with successful
switchover (note that if the UPF needs to be changed, the SMF will first select the
UPF and update the information of the appropriate UPF), if there is failed QoS
Flow established among them, the SMF will initiate the PDU session modification
process. For the failed switches, SMF will execute the PDU session release or
deactivation process according to specific reasons.

The N2 switching process is shown in figure 3-18.

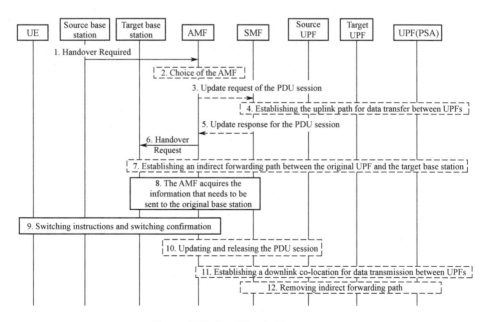

Figure 3-18 The N2 switching process

PREPARATION PHASE

Similar to the switching of Xn interfaces, before the switching step is executed, the
source base station, target base station, and Core Network need to be prepared for
the switching, and the related resources after the switching need to be prepared,
specifically the following: (1) establishing the user-plane resources for uplink

transmission after the switchover, and (2) establishing the downlink forwarding tunnel.

The main flow of the preparation phase is as follows:

(1) The source base station sends a Handover Required message to the source AMF network element, which contains the identification of the target base station, the PDU session information that needs to be switched, and the container information sent to the target base station.

(2) The AMF selects whether the AMF can continue to provide service to the UE based on the identification information of the target base station. If the service cannot be continued, the AMF (S-AMF) selects a new AMF (T-AMF), and the S-AMF includes the contents of the source base station message and the UE context information in the UE context establishment request message and sends it to the T-AMF. Thereafter, the T-AMF establishes the association with the corresponding PCF according to the message.

(3) The AMF sends a message to the corresponding SMF to update the corresponding PDU session according to the PDU session information that needs to be switched, combined with the slices it can serve.

(4) The SMF confirms whether the corresponding PDU session can be switched, and at the same time, the SMF checks whether a new UPF needs to be inserted according to the location of the UE, after which the SMF establishes the uplink between the UPFs.

(5) The SMF sends the relevant N2 SM information to the AMF according to the success or failure of the PDU session establishment, or the failure reason value.

(6) Thereafter, the AMF sends the container and N2 MM/SM information sent by the source base station to the target base station via Handover Request. The target base station establishes the corresponding connection resources for the corresponding PDU session and returns the corresponding N3 information of the PDU session on the RAN side. For the successfully established PDU session, the uplink transmission path between RAN-UPF is successfully established.

(7) If the indirect forwarding path needs to be established, the SMF and the corresponding UPF need to interact separately to establish the indirect forwarding path from S-UPF to T-RAN. The path from S-RAN to S-UPF will be established during the execution.

(8) Thereafter, the AMF obtains the information that needs to be sent to the source base station, including the information of the PDU session establishment, the information of the S-UPF used for forwarding, etc.

EXECUTION PHASE

The main flow of the execution phase is as follows:

(1) After receiving the information from AMF about switching (e.g., PDU session information), the source BTS instructs the UE to perform switching. After the synchronization between the UE and the new cell is completed, the UE sends a switching confirmation message to the target base station to complete the switching of the air interface. Thereafter, the base station will inform the AMF that the switchover is successful.

(2) If the T-AMF cannot support some PDU sessions due to slicing, then the T-AMF will trigger the PDU session release process. For other PDU sessions, the T-AMF will update the information of the PDU session at the SMF.

(3) The SMF interacts with the corresponding UPF to establish the path for downlink data transmission.

(4) The SMF removes the corresponding indirect forwarding tunnel.

3.2.1.5.3 Service request process

The main purpose of the service request process is to activate a PDU session for a connected UE, or to change an idle UE to the connected state and optionally activate a PDU session to enable receiving downlink/uploading uplink data or signal. As for sessions initiated by the network side and the UE side, the service request process can be divided further in order to meet both needs. The service request process initiated by the UE side is demonstrated in figure 3-19.

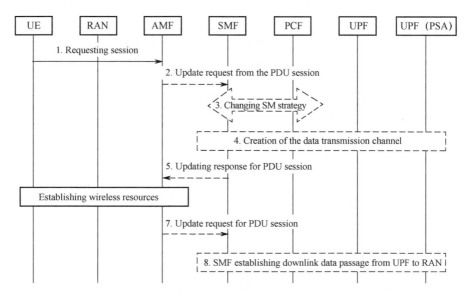

Figure 3-19 The service request process initiated from the UE side

Step 1: The UE sends a service request message to the AMF. The message contains required parameters like the PDU session identifier needing activation.

Step 2: The AMF invokes the session service of the selected SMF to trigger a session update based on the mapping relationship between the locally stored session identifier and the SMF.

Step 3: The session management policy connection between SMF and PCF is updated, and the session policy rules are obtained based on the PCF's decision. The session policy control is detailed in section 3.2.4 of this book.

Step 4: SMF regulates UPF in establishing the data transportation channel. SMF decides whether an indirect data forwarding channel is needed based on whether the old UPF needs to be deleted.

Step 5: SMF returns the response message of the service session.

Step 6: The AMF sends a N2 request to the RAN and requests RAN to establish wireless resources. The RAN thereafter establishes the DRB corresponding to the QoS Flow via the RRC reconfiguration message.

Step 7: After obtaining the return message from the RAN, for each PDU session, the AMF notifies the corresponding SMF via the service session of the received/ rejected QoS Flow information and the N3 tunnel information of the RAN.

Step 8: The SMF establishes the downlink data path from the UPF to the RAN.

The network-side triggered service request flow is shown in figure 3-20.

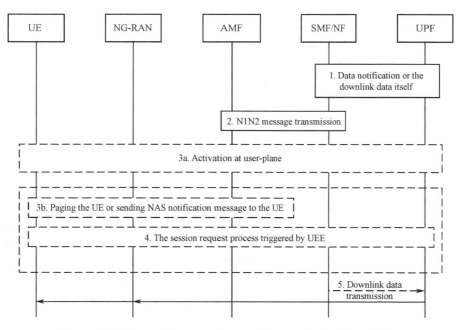

Figure 3-20 The service request process triggered by the Internet side

Step 1: When the UPF receives downlink data, depending on the instructions of the SMF, the UPF may locally cache the downlink data and send a data notification message to the SMF, or the UPF may send the downlink data directly to the SMF.

Step 2: When the SMF receives the downlink data or the data notification message sent by the UPF, the SMF finds the PDU session of the corresponding data and the QoS Flow information; thereafter, the SMF invokes the N1N2 message transmission request service provided by the AMF including information like the QoS information and tunnel information. If the service request flow is triggered by the Core Network elements (e.g., SMF, LMF, and SMSF) needing to establish NAS connection with UEs or send N1 messages to the UEs, then the requested service for N1N2 message transmission will contain N1 messages.

Step 3a: If at this time the AMF receives PDU session identification from the SMF and the UE is in the connected state for the access type corresponding to this PDU session, then the AMF will initiate the user-plane activation process as described in steps 12 to 22 of section 4.2.3.3 of 3GPP TS 23.502 [2].

Step 3b: If at this time the AMF receives the identification of the PDU session from the SMF and the UE is in the idle state for the access type corresponding to this PDU session, then the AMF may initiate a paging process for the UE. For the specific process of paging, see the description of steps 3 to 4 in 3GPP TS 23.502 [2]; if the AMF finds that the UE is in another access-type connected state, then the AMF may inform the UE to initiate the service request process via a NAS notification message.

Step 4: The UE initiates the service request process triggered by the UE after receiving the appropriate message from step 3b.

3.2.1.5.4 The Access Network release process

The Access Network release process is used to release the signaling connection between the UE and AN, the N2 connection between the AN and the AMF, and the N3 connection between the AN and the UPF. This process may either be:

(1) triggered by the AMF. Due to an unknown error in the AMF, or if the UE is in the idle state for registration and does not need to activate the user-plane (when the AMF may decide to change the UE to the idle state), or if the timer associated with the UE registration update stored locally in the AMF expires;

(2) triggered by NG-RAN. Possible causes include an unknown error in the NG-RAN, wireless connection failure, user-plane deactivation, or mobility restrictions.

The Access Network release process is shown in figure 3-21.

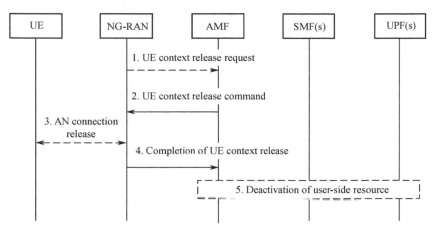

Figure 3-21 The Access Network release process

Step 1: When the NG-RAN confirms the need to release the UE context, the NG-RAN sends a UE context release request to AMF containing the reason value and the ID of the PDU session.

Step 2: The AMF sends the UE context release message to NG-RAN containing the reason value. If this is a release initiated by the NG-RAN, then the reason value stays the same as step 1.

Step 3: If the AN connection between the UE and the NG-RAN still exists now (e.g., an AN release by an AMF decision), then the NG-RAN releases the AN connection with the UE at this time.

Step 4: The NG-RAN confirms the release by sending a UE Context Release Completion message to the AMF. Depending on the specific reason for the release, the message can contain the PDU session ID and the UE location information.

Step 5: Based on the received PDU session ID, the AMF interacts with the corresponding SMF in order to deactivate the corresponding user-plane resources.

3.2.1.5.5 Deregistration process

The deregistration process is used when the UE informs the network that it no longer needs access to the 5G system, or when the network informs the UE that the UE cannot access the network again. Specific reasons for this include UE shutdown and the UE contract information change. Deregistration processes can be divided into two categories: UE-initiated deregistration processes and network-initiated deregistration processes.

The deregistration process initiated by the UE is shown in figure 3-22.

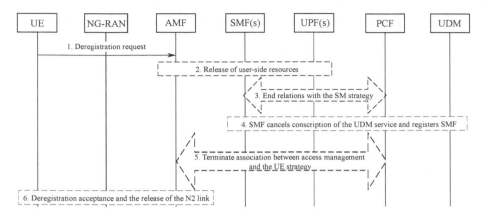

Figure 3-22 The deregistration process initiated by the UE

Step 1: The UE sends a deregistration request NAS message to the AMF, which contains information such as the 5G-GUTI of the UE, the reason value for deregistration, and the access type corresponding to the deregistration.

Step 2: If the UE establishes a PDU session in the network corresponding to this access type, then the AMF invokes the service used by the corresponding SMF to release the PDU session. Thereafter, the SMF and the UPF inter-release the corresponding context information.

Step 3: If there are dynamic Policy and Charging Control (PCC) rules for the PDU session, then the SMF also interacts with the PCF to terminate the association between the SMF and the PCF to the SM policy.

Step 4: The SMF invokes the UDM's unsubscribe and deregistration service to remove the information about the SMF at the UDM.

Step 5: If no other access type exists for this UE other than the access type to be deregistered, then the AMF also interacts with the PCF to terminate the association of the AMF with the PCF associating with the UE and the mobile management policy respectively.

Step 6: Depending on the deregistration type, the AMF may send a deregistration acceptance message to the UE. If a corresponding N2 connection exists, the AMF will release this N2 connection.

The network-initiated deregistration process is shown in figure 3-23.

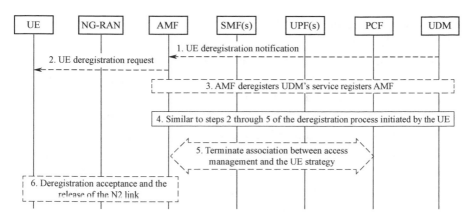

Figure 3-23 The deregistration process initiated by the network

Step 1: If the UDM needs to delete the registration information and PDU session of the UE, the UDM informs the AMF via a deregistration notification message, which contains information such as the reason value for deregistration, and the access type for deregistration.

Step 2: The AMF sends a deregistration request message to the UE.

Step 3: The AMF invokes the UDM's unsubscribe and deregistration services in order to remove the AMF-related information from the UDM.

Step 4: The interaction between the AMF, SMF, UDM, and PCF thereafter is similar to steps 2 to 5 of the deregistration process initiated by the UE.

Step 5: If no other access type exists for the UE other than the access type to be deregistered, then the AMF also interacts with the PCF separately to terminate the association between the AMF and the PCF for the UE and the mobility management policy.

Step 6: Depending on the deregistration type, the AMF may send a deregistration acceptance message to the UE. If a corresponding N2 connection exists, the AMF releases this N2 connection.

3.2.1.6 Summary

Mobility management is the basic mechanism used by 5G networks to maintain user locations for data transmission in response to the mobile and power-constrained nature of UEs. In addition to the connection state management, registration state management and the corresponding transition process already supported by 4G networks, 5G networks can support more diverse mobility management for UEs with different characteristics by introducing features such as RRC Inactive state, MICO mode, and on-demand mobility management.

3.2.2 Session management

3.2.2.1 Introduction

The total transmission resources of the network are limited due to restrictions such as the air interface spectrum and network bandwidth. In order to efficiently and properly secure the network's transmission capability for different user service flows through the limited transmission resources, the wireless network needs the ability to control each data transmission connection from the user to the network (including lifecycle control like establishing connections, allocating and adjusting resources, and release), so as to use the network efficiently to transmit resources. In 5G networks, this user-to-network connection relationship is referred to as a PDU (Protocol Data Unit) session.

A PDU session is an abstract concept that represents the logical connection between the UE and the DN through AN and UPF. Each PU session is allocated with an independent session identifier, and the 5G network regulates the creation and release process of a PDU session.

In order to communicate through the 5G network and data network, the UE first needs to establish and initiate one or more PDU sessions. During this process, the network will allocate user-plane transmission resources for that PDU session, including establishing the air interface resources and Core Network tunnel transmission resources for the respective PDU, in order to open the transmission tunnel between the UE and the UPF. Afterwards, the user message table can be transmitted between the UE and the data network through the respective user-plane transmission resources of the session and can implement QoS policy control at the flow granularity.

In general, a PDU session of each IP type can be allocated with an IPv4 address or an IPv6 prefix, but in certain bypass scenarios, it is possible for a PDU session to be allocated with multiple IPv6 prefixes.

The concept of PDU-like sessions in 4G networks is called PDN (Packet Data Network) connectivity. Compared with PDN connections in 4G networks, PDU sessions in 5G support more session types to meet the session management needs of more scenarios, including vertical industry services. Specifically, the 5G network supports the following five types of PDU sessions:

(1) IPv4: to carry IPv4 protocol messages
(2) IPv6: to carry IPv6 protocol messages
(3) IPv4v6: to carry IPv4 or IPv6 protocol messages
(4) Ethernet: to carry Ethernet frame messages
(5) Unstructured: to carry unstructured messages, including non-standardized protocols or, for 5G networks, protocol messages of unknown protocols

3.2.2.2 User-plane protocol stack design for sessions

The user-plane protocol stack used for PDU session transmission in 5G is shown in figure 3-24. Messages in a PDU session in a 5G network need to comply with this protocol stack during end-to-end transmission. Depending on the AN type, different Access Network protocol stacks can be used between the UE and the AN. The user-plane of the Core Network between the AN and the UPF and between the UPF and the UPF uses the GTP-U protocol for tunnel forwarding. The protocol layer between the UE and the Access Network and the GTP-U protocol layer of the Core Network carry the PDU layer from the UE to the anchor UPF above. Depending on the PDU session type, the specific implementation of the PDU layer can be IPv4 protocol messages, IPv6 protocol messages, Ethernet frames, or unstructured user messages. Depending on the deployment, one or more Intermediate UPFs (I-UPF) may exist between the 5G-AN and the anchor UPF.

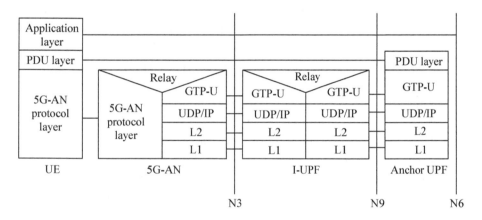

Figure 3-24 5G User-plane protocol stack

3.2.2.3 Session continuity guarantee

In 4G networks, PDN connections are always anchored to a specific PDN GW. During the lifecycle of PDN connections, the corresponding IP address or prefix stays constant to keep the session coherent. In order to meet the continuity needs of different services, the 5G network is designed with three different Session and Service Continuity mode (SSC mode). Three different SSC modes are shown in figure 3-25.

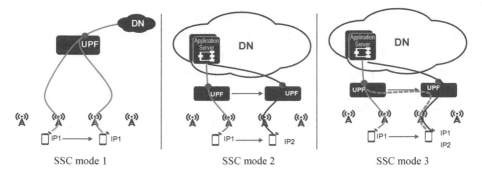

Figure 3-25 SSC mode in different scenarios

SSC mode 1: The PDU session in this mode provides session continuity. Similar to 4G PDN connections, the UPF, which is the user-plane anchor point, remains unchanged during the session lifetime. For IP type PDU sessions, the IP address or prefix of the UE also remains unchanged. In general, the UPF of the anchor point for this type of PDU session is deployed in a more centralized location and is mainly used for typical services with high continuity requirements such as IMS voice.

SSC mode 2: The PDU session in this mode does not guarantee service continuity for the UE. When the network determines that the anchor UPF needs to be changed (e.g., the routing path of the anchor UPF is too long relative to the current location of the UE), the network will trigger the release of the current PDU session and instruct the UE to establish a new PDU session and select the new anchor UPF in the process of establishing the new PDU session, so that the routing path of the new session is more optimized. This way of establishing a PDU session is also called "Break-before-Make." This type of PDU session is mainly for web browsing, video-on-demand with caching capability, and other typical services that allow brief connection interruptions.

SSC mode 3: The PDU session in this mode is able to provide service continuity services to the UE, but not session continuity. Unlike SSC mode 1, the anchor UPF of the service flow can be changed. The network instructs the UE to establish a new PDU session and selects the local anchor UPF with more optimized routing during the establishment of the new PDU session; after the new PDU session is established for a certain time, the old PDU session is released. In this process, although the IP address/prefix of the UE will change, the service continuity can still be maintained because the UE always has at least one PDU session accessing the DN at any moment. This way of establishing PDU sessions is also called "Make-before-Break."

The SSC mode of a PDU session stays constant during the lifecycle of the session. The PDU mode employed by each session is decided collectively by UE's request, the user's contracting strategy, and the carrier's strategy. The US has the SSC mode choice strategy configuration provided by the carrier. This SSC mod choice strategy is a part of the URSP (UE Route Selection Policy) (see section 3.2.4.4.2), which includes the application and its associate SSC mode. The UE will decide the SSC mode accordingly when an application is launched. If the current PDU session of the UE can satisfy the needs of the said application, the UE can use the current PDU session to transmit its data message. Otherwise, the UE will initiate the session initiation process and carry the requested SSC mode in order to satisfy the needs of the said application. The contract data acquired by the SMF from the UDM include a list of allowed SSC modes and default SSC modes for the DNN and the S-NSSAI (Single Network Slice Selection Assistance Information). When the SMF receives a session creation request from the UE, the SMF decides whether it can accept the request based on the information listed above. When the SMF declines the said request, the SMF sends the result value and the allowed SSC mode to the UE. The UE may attempt to create a PDU session gain based on the received result value, the allowed SSC modes, and other rules of the SSC mode choice strategy. When the session request of the UE does not contain a chosen SSC mode, the SMF will choose the default SSC mode to be the SSC mode of that PDU session.

3.2.2.4 Ethernet PDU session vs. Unstructured PDU session
The 5G network provided the ability to transmit data messages of the Ethernet PDU session type and the unstructured PDU session type in order to support service for vertical industries including industrial networks.

For the Ethernet-type PDU session, the SMF and the UPF anchor need to be expanded to support the following functions to allow the transmission of Ethernet frames:

(1) The anchor UPF detects the source address, i.e., the UE MAC address, in the uplink Ethernet frame sent by the UE.
(2) The SMF indicates the UE MAC address used in the UPF uplink PDU session.
(3) If the anchor UPF receives a downlink ARP/IPv6 Neighbor Solicitation request, in order to avoid broadcasting the message to all UEs, the anchor UPF may either answer the above request message based on local cache information (i.e., mapping the relationship between UE MAC and the UE IP) or forward the ARP message to the SMF for processing.

The forwarding method of Ethernet frames at the N6 interface is decided by the configuration of the carrier:

(1) When the anchor UPF receives a downlink Ethernet frame, if the PDU session has a one-to-one correspondence with the N6 interface, the N6 interface can be implemented through a dedicated tunnel. In this scenario, the anchor UPF just forward Ethernet frames without sensing the UE MAC address.

(2) When the anchor UPF receives a downlink Ethernet frame, if the PDU session corresponds to the N6 interface in a many-to-one relationship, the anchor UPF needs to sense the UE MAC address in order to determine the PDU session corresponding to the downlink Ethernet frame and send the Ethernet frame to the UE through this PDU session.

For the unstructured type of PDU sessions, the N6 interface can use point-to-point tunneling to transfer data. Taking the use of a point-to-point tunnel based on UDP/IPv6 encapsulation as an example, the process of achieving data transmission for PDU sessions of unstructured session type is as follows:

(1) The SMF assigns an IPv6 prefix to the PDU session. The IPv6 prefix is only used to perform tunnel forwarding on the N6 interface and to identify the PDU session corresponding to the downlink message, and thus does not need to notify the UE.

(2) The UPF acts as a transparent forwarding node between the UE and the DN. For the uplink unstructured message, the anchor UPF encapsulates the IPv6 prefix for it and forward it to the DN through the N6 tunnel. For the downlink unstructured message, the destination IP is the IPv6 prefix corresponding to the PDU session, and the UDP port number is the UDP port number defined by 3GPP. The anchor UPF decapsulates the received message, determines the session corresponding to this downlink message according to the IPv6 prefix, and then forward it to the UE through this PDU session.

The selection of the session type is similar to the selection method of the SSC mode described earlier. The URSP (or local configuration policy) provided by the operator to the UE contains the session type selection policy corresponding to the application (see section 3.2.4.3 for details). When the application is started, the UE determines the session type corresponding to the application according to this session type selection policy. Based on the session type corresponding to the application, the UE determines whether there already exists an existing PDU session that can satisfy the session type of the application. If it exists, the UE can reuse the existing PDU session to transmit the data message of the application;

otherwise, the UE initiates the session establishment process and carries the requested session type to establish a new PDU session that satisfies the above application session type. The SMF determines whether to accept the requested session type based on the local policy and the contract information of the UE.

3.2.2.5 Key processes for session management

3.2.2.5.1 Session creation process

The session creation process is employed by the UE for the creation of a new PDU session and for distributing the user-plane connection resources between the UE and UPF anchor point. The PDU creation process is shown in figure 3-26. Due to the design separating mobility management and session management, the UE does not directly finish the creation of the session upon initial registration to the network in the 5G network but does so per request after the UE finishes the registration. When an application initiates a network connection, the UE decides whether to initiate the following session creation process to establish end-to-end transmission resources according to the corresponding URSP rules of the application or local configurations.

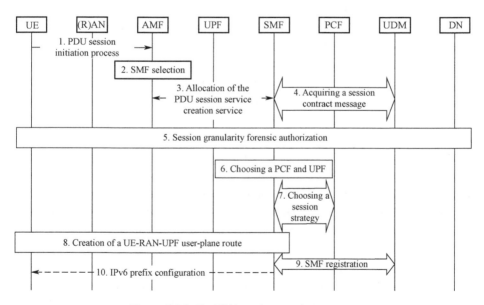

Figure 3-26 The PDU session creation process

Step 1: The UE sends the session creation message to the AMF. This message contains required parameters such as session identification, type of session, SSC mode, DNN, and the S-NSSAI. The UE can determine some of the request parameters in the above session creation request message based on the application of the corresponding URSP rules.

Step 2: SMF selection. The AMF selects the appropriate SMF for the session based on the DNN, S-NSSAI, and contracted data. For multiple session requests established by the UE in the same DN and slice, the AMF tries to ensure the selection of the same SMF to reduce the number of SMFs serving the same UE.

Step 3: The AMF invokes the session service of the selected SMF to trigger session creation.

Step 4: The SMF obtains the session management signing data from the UDM, including information on the user's permitted SSC mode, session type, default session parameter values, and the signed Session Aggregate Maximum Bit Rate (Session-AMBR).

Step 5: This step is optional. The SMF and the third-party authentication, authorization, and accounting server, i.e., Authentication, Authorization, and Accounting (AAA), in the DN perform secondary authentication and authorization of the session. See section 3.2.2.5.5 for details on secondary session authentication and authorization.

Step 6: The SMF selects the PCF and UPF for the session.

Step 7: A session management policy connection is established between the SMF and the PCF, and session policy rules are obtained. See section 3.2.4.6 for details.

Step 8: The SMF establishes an end-to-end user-plane connection between the UE, AN, and UPF. In this process, the SMF sends a session creation acceptance message to the UE through the AMF and AN, which contains the selected SSC mode, selected session type, S-NSSAI, and DNN.

Step 9: The SMF registers with the UDM, and the UDM records the SMF identification corresponding to the session.

Step 10: The SMF assigns an IPv6 prefix to the UE. When the session type is IPv6 or IPv4v6, the SMF generates an RA message carrying the IPv6 prefix assigned for the session and sends it to the UE via the user plane.

3.2.2.5.2 Session modification process

The session modification process is shown in figure 3-27. The modification of the PDN connection in 4G is carried out according to the bearer granularity, including proprietary bearer activation, bearer modification, and the bearer deactivation process. Unlike 4G, the session modification in 5G is carried out according to the QoS Flow granularity. The network can execute the increase, edit or deletion of the QoS Flow. The relationship between QoS Flow and the session is described in section 3.2.3.

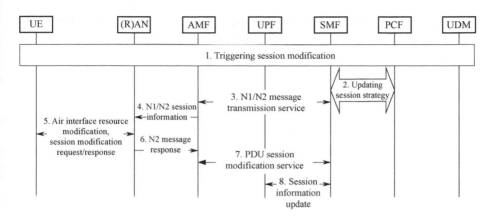

Figure 3-27 The session modification process

Step 1: The modification of a PDU session may be triggered by a number of different events, including

(1) UE triggers, such as a UE request to add, modify, or delete QoS Flow;
(2) PCF triggers, such as a PCF initiating a policy update based on internal or external state modifications;
(3) UDM triggers, such as the update of session-related contract data;
(4) SMF triggers, such as SMF triggers to add, modify, or delete QoS Flow based on local policy;
(5) AN triggers, such as when the AN determines that the QoS characteristics of certain QoS Flows cannot be satisfied or can be satisfied again, the AN can also notify the network by initiating a session modification process.

Step 2: If the session modification causes the SMF to need to re-request the session policy, the update of the session management policy is performed between the SMF and the PCF.

Step 3: The SMF invokes the N1/N2 messaging service of the AMF to send the updated session information of the N1 and/or N2 interfaces to the AMF, where the N1 session information contains the QoS rules sent to the UE and the N2 session information contains the QoS configuration documents sent to the AN. It is important to note that message names used in step 3 may differ depending on the trigger conditions of the session modification process.

Step 4: The AMF sends the N1 and/or N2 information obtained from the SMF to the AN via the N2 message.

Step 5: The AN initiates the air interface resource modification procedures based on the received updated QoS parameters and updates the air interface resources involved in this session modification. At the same time, if the AN receives an N1 message (e.g., PDU session modification indication or answer message) from the SMF, the AN sends this N1 message to the UE.

Step 6: The AN sends an N2 answer message to the AMF containing the list of accepted QFIs and/or the list of rejected QFIs.

Step 7: The AMF invokes the session update service of the SMF and sends the information obtained from the AN to the SMF.

Step 8: If required, the SMF updates the new session information (e.g., updated QoS parameters) to the UPF via the N4 interface.

3.2.2.5.3 Session release process

The PDU session release process is used to release all the transmission resources associated with the session, as shown in figure 3-28.

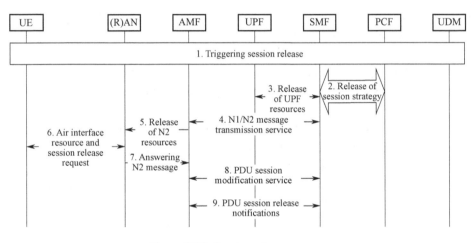

Figure 3-28 Session release process

Step 1: Similar to the PDU session modification process, the release of the PDU session may also be triggered by a number of different events, including

(1) the UE is initiating a PDU session release;

(2) PDU session release initiated by PCF according to the policy;

(3) AMF initiation. AMF can be triggered when the session state maintained on the UE is different from the session state maintained on the AMF;

(4) AN initiation. The AN initiates the PDU session release process when all the resources associated with the PDU session on the AN are released, for example, when all the QoS Flows of the PDU session are released; and

(5) SMF initiation. For example, when the SMF receives an authorization termination request from DN-AAA, a signing update request from UDM, or a local configuration policy trigger, the SMF will initiate the PDU session release process.

Step 2: If required, the SMF initiates a session management policy release with the PCF.

Step 3: The SMF releases the session resources on the UPF through the N4 interface.

Step 4: The SMF invokes the N1/N2 messaging service of the AMF to send the session release message of the N1/N2 interface to the AMF.

Step 5: The AMF sends the N1/N2 information obtained from the SMF to the AN via the N2 message.

Step 6: The AN sends the session release request message to the UE and releases the air interface resources allocated for the PDU session.

Step 7: The AN sends an N2 response message to the AMF.

Step 8: The AMF calls the PDU session update service to send the answer message received from the AN to the SMF.

Step 9: The SMF sends a PDU session release notification event to the AMF to release the session-related binding relationship on the AMF.

3.2.2.5.4 Selective deactivation process of the PDU session

The selective activation/deactivation of a PDU session is a new feature of 5G. For an established PDU session, the 5G network allows the user-plane connection of the said PDU session to be selectively deactivated, which is to release the N3 tunnel connection and air interface resources of the said session but meanwhile retaining control-plane signaling connections between network elements such as AMF, SMF, and PCF.

The selective deactivation session for a PDU session is shown in figure 3-29.

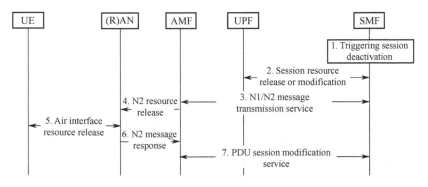

Figure 3-29 The selective deactivation process of the PDU session

Step 1: The deactivation of a PDU session may be triggered by the following events:

(1) During the switching process, the target AN rejects all QoS Flow for a PDU session, resulting in the failure of the switching of this PDU session.
(2) For a PDU session accessing the LADN, the AMF sends a notification to the SMF that the UE has moved out of the LADN service area.
(3) The AMF sends a notification to the SMF informing the UE that it has moved out of the Allowed Area.

Step 2: The SMF releases (if the N3 UPF is still reserved) or modifies the session resources on the UPF via the N4 interface.

Step 3: The SMF calls the N1/N2 messaging service of the AMF and sends the session release message of the N2 interface to the AMF.

Step 4: The AMF sends the N2 information obtained from the SMF to the AN via N2 message.

Step 5: The AN releases the air interface resource corresponding to the session.

Step 6: The AN sends an N2 answer message to the AMF.

Step 7: The AMF invokes the PDU session update service to send the answer message received from the AN to the SMF.

For deactivated sessions, the PDU session can be activated through a service request process initiated by the UE or the network. This process is similar to the UE or network-initiated service request process in 4G. The difference is that in 5G it is possible to selectively activate one or several PDU sessions, while in 4G, only the user-plane connections of all PDN connections of the UE can be activated. See section 3.2.1.5.3 for the specific process.

3.2.2.5.5 Secondary authentication process for PDU sessions

During the PDU session creation process, the identity of the UE in the DN network's PDU session granularity authorization can be done by a DN-AAA deployed by the carrier or a third party. The SMF determines whether to authorize the creation of the PDU session based on local strategy. If the said authorization fails, the SMF refuses the PDU session creation process. The secondary authorization process of a PDU session is shown in figure 3-30.

Step 1: The SMF chooses the UPF per request and creates the route between the SMF and the DN-AAA. This UPF is only used for forwarding messages between the SMF and the DN-AAA and can be different from the anchor UPF of the PDU session. It is also possible for the SMF to communicate with the DN-AAA directly if the DN-AAA is in the 5G network.

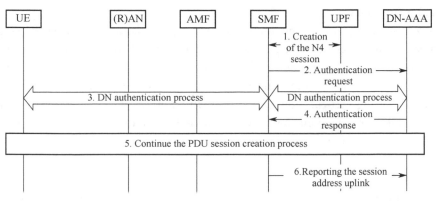

Figure 3-30 The secondary authorization process of a PDU session

Step 2: The SMF sends the authorization/authentication request to the DN-AAA.

Step 3: The DN-AAA runs the DN authentication process with the UE through the SMF. The SMF transmits these authentication messages between the UE and the SMF by NAS messages through invoking the N1/N2 message transmission service of the AMF.

Step 4: After successful authentication, the DN-AAA provides the DN Authorization Data and the IP address/IPv6 prefix of the PDU session to the SMF. The authentication data contains at least one of the following elements:

(1) DN Authorization Profile Index for locally configured policy and billing control information on the SMF or PCF

(2) For Ethernet sessions, may also include a list of MAC addresses allowed for the session or the allowed VLAN flags

(3) DN Authorization Session Aggregate Maximum Bit Rate (AMBR)

Step 5: The SMF continues the PDU session creation process.

Step 6: After the completion of the session creation process, the SMF can inform the DN-AAA of information like the IP address of the session, the N6 routing information, and the UE MAC address, to the DN-AAA based on its subscription instructions.

The DN-AAA can initiate the reauthorization, cancellation, or update of the PDU session at any time. The SMF can release or update the PDU session per request of the DN-AAA.

3.2.2.6 Summary
Session management is the basic mechanism for controlling user message transmission in 5G networks. 5G session management inherits the 4G session management mechanism, and enhances the ability to support multiple session types, services, and session continuity modes to support more diverse service types, including vertical industry services. In addition, 5G session management and mobility management have been designed separately to support the more flexible deployment of network element functions.

3.2.3 QoS
3.2.3.1 Introduction
Quality of Service (QoS) describes a set of service requirements a network must meet to ensure the appropriate level of service to secure data transmission. 5G networks need to support the transmission of a wide variety of differentiated services, such as video, mobile payments, web browsing, and factory automation control, all of which require different qualities of service. For example, video services require larger bandwidth, while automation and control services generally require lower lag and higher reliability. Based on the QoS framework of the 5G networks, operators can provide different QoS guarantees for each service.

This section explains the QoS framework for the 5G system architecture, including the division of labor for QoS control between the UE, Access Network, and Core Network, and the introduction of QoS parameters.

3.2.3.2 QoS framework
In 5G networks, a QoS Flow is the finest granularity for end-to-end QoS control in a PDU session. Each QoS Flow uses a QoS Flow Identifier (QFI), which is unique within a PDU session. Within a PDU session, user-plane packets with the same QFI are handled in the same manner (e.g., scheduling, admission thresholds) during transmission.

The QoS framework of the 5G network is shown in figure 3-31.

In 5G, each UE can create one or more PDU sessions, and at least one QoS Flow exists in each PDU session.

When the 3GPP is connected to the network, the NG-RAN establishes one or more radio bearers for each PDU session, including a default radio bearer. Each radio bearer can only serve one PDU session. The wireless bearer is the smallest granularity of packet processing for the air interface. Each wireless bearer provides the same processing for message forwarding. The NG-RAN can establish separate wireless bearers for QoS Flows, requiring different message-forwarding processing, or multiple QoS Flows belonging to the same PDU session can be bound to the same wireless bearer.

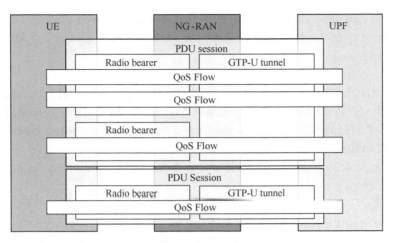

Figure 3-31 The QoS framework of the 5G network

For unstructured type PDU sessions, only one QoS Flow is supported to be established per unstructured type PDU session.

Each QoS Flow is associated with the following three types of information for end-to-end QoS control:

(1) The QoS Profile provided by the SMF to the AN
(2) One or more QoS Rules and QoS parameters at the QoS Flow level provided by the SMF to the UE
(3) One or more Packet Detection Rules (PDR) for uplink and downlink messages and the corresponding QoS Enforcement Rule (QER) provided by the SMF to the UPF

The end-to-end QoS control and mapping relationship of QoS Flow at the user-plane nodes is shown in figure 3-32.

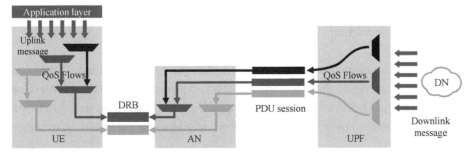

Figure 3-32 The end-to-end QoS control and mapping relationships at user-plane nodes of the QoS Flow

In the downlink direction, the UPF matches the received packets according to the priority of the downlink packet filter set in the PDR sent by the SMF from highest to lowest:

(1) If a matching downlink PDR is found, the corresponding QFI is encapsulated in the GTP-U header of the message according to the matching result, and the NG-RAN maps the packet to the corresponding wireless bearer based on the QFI in the GTP-U header.

(2) If no matching downlink PDR is found, the UPF will discard the downlink packet.

In the uplink direction, the UE matches the packets to be sent according to the uplink packet filter set in the QoS Rule according to the priority of the QoS Rule from highest to lowest:

(1) If there is a matching QoS Rule, the UE binds the uplink packet to the QoS Flow using the QFI in the corresponding QoS Rule, then binds the QoS Flow to the corresponding wireless resource (i.e., wireless bearer).

(2) If there is no matching QoS Rule, the UE will discard the uplink packet.

3.2.3.3 Key QoS mechanisms

3.2.3.3.1 QoS control process

Similar to 4G networks, 5G networks can perform the management flow of QoS control through the signaling plane.

Using the establishment of QoS Flow as an example, the SMF determines the establishment of QoS Flow based on the local policy or PCC rules sent by the PCF. The specific flow is as follows:

(1) The SMF sends PDR and QER to the UPF.

(2) The SMF sends the QoS Profile of the QoS Flow to the AN through the AMF.

(3) SMF sends QoS Rule to the UE through AMF and AN, which contains QoS control information.

Through the above signaling interaction, the QoS Flow is established between the UE, AN, and UPF. the AN allocates the radio resources of the air interface according to the QoS Profile and stores the binding relationship between the QoS Flow and the radio resources. The QoS Flow creation process is shown in figure 3-33.

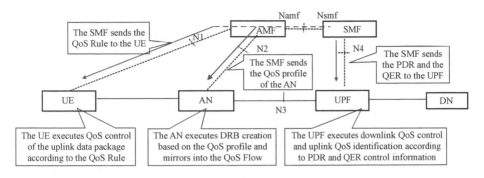

Figure 3-33 QoS Flow creation process

Further, when there is a change in the established QoS Flow, such as a modification or deletion of the QoS Flow, the SMF updates the QoS information on the UE, AN, and UPF through the PDU session modification flow (see section 3.2.2.5.2).

3.2.3.3.2 Reflective QoS Control (RQC)

Reflective QoS Control means that the UE can generate its own QoS Rule for the corresponding uplink packet based on the packets sent downlink from the network and use the corresponding QoS Rule to perform QoS control of the uplink data. The main purpose of introducing the reflective QoS control mechanism is to avoid the signaling overhead between SMF and the UE caused by the frequent update of QoS Rule.

The reflective QoS feature is applicable to both IP-type PDU sessions and Ethernet-type PDU sessions, and the UE instructs the SMF whether the reflective QoS mechanism is supported or not in the PDU session creation process. The reflection QoS control process is shown in figure 3-34.

When a PCF is deployed in a 5G network, the PCF determines whether a particular Service Data Flow (SDF) uses reflected QoS control. Otherwise, the SMF determines whether the SDF uses reflected QoS control. Specifically, using the scenario with PCF deployed as an example, in figure 3-34, when Reflective QoS is used for a specific SDF, the PCF includes Reflective QoS Control (RQC) in the generated PCC rules to indicate the service to use Reflective QoS Control. Further, the SMF determines that the SDF information provided to the UPF contains Reflective QoS Indication (RQI) based on the received RQC and includes Reflective QoS Attribute (RQA) in the QoS Profile sent to the NG-RAN.

For each SDF corresponding to a downlink packet, the UPF sets the RQI identifier in the encapsulation header of the GTP-U. When a downlink packet received by the NG-RAN contains RQI, the QFI and RQI of the downlink packet

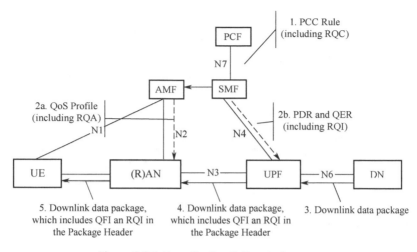

Figure 3-34 The reflective QoS control process

are indicated to the UE. When a downlink data containing RQI is received on the UE side, then:

(1) if the packet filter corresponding to the downlink packet does not exist in the QoS Rule pushed by the UE, then the UE creates a new pushed QoS Rule with the packet filter corresponding to the downlink packet in the QoS Rule, and the UE starts a timer for the pushed QoS Rule; or

(2) if a packet filter corresponding to the downlink packet exists in the QoS Rule pushed by the UE, the UE restarts the timer corresponding to the pushed QoS Rule. Further, if the QFI corresponding to said downlink packet is different from the pushed QoS Rule, the UE updates the QFI of the pushed QoS Rule.

When the network side decides to no longer use RQC for a certain SDF, the SMF deletes the RQI from the SDF message provided to the UPF. The UPF does not set the RQI in the package heading of the GTP-U, and for a duration of time, the UPF keeps receiving the corresponding uplink service of the SDF. The UE deletes the QoS Rule deduced by the UE according to the timer.

3.2.3.4 QoS parameters
3.2.3.4.1 QoS parameters for QoS Flow granularity
The 5G QoS frame supports QoS Flows that require a guaranteed flow bit rate (GBR QoS Flow) and QoS Flows that do not require a guaranteed flow bit rate (Non-GBR QoS Flow). For GBR QoS Flows, the network needs to reserve resources in order to guarantee its bandwidth. For Non-GBR QoS Flows, such reservations are not required.

The QoS control of the QoS Flow is determined by the related parameters. For the QoS parameters of different types of QoS Flow see table 3-4.

Table 3-4 QoS parameters of different QoS types

QoS Flow type	QoS parameters
Non-GBR QoS Flow	5QI, ARP, RQA (optional)
GBR QoS Flow	5QI, ARP, GFBR, MFBR, Notification Control (optional), Maximum Packet Loss Rate (optional)

Note:
(1) 5QI: 5G QoS Identifier
(2) ARP: Allocation and Retention Priority
(3) RQA: Reflective QoS Attribution
(4) GFBR: Guaranteed Flow Bit Rate
(5) MFBR: Maximum Flow Bit Rate
(6) Notification Control
(7) Maximum Packet Loss Rate

The meaning and the major function of each QoS parameter are as follows:

(1) 5QI is an index identification of the QoS feature similar to the QoS Class Identifier (QCI) in 4G. It is used to control the QoS forwarding processing of QoS Flow (like Scheduling weights, admission thresholds, queue management thresholds, and link layer protocol configuration.). 5QI is divided into three main categories: standardized 5QI, pre-configured 5QI, and dynamically assigned 5QI. Standardized 5QI values correspond to a standardized combination of 5G QoS features specified in the standard; pre-configured 5QI values correspond to 5G QoS features that are pre-configured in the AN, and the meaning of pre-configured 5QI can vary due to pre-configurations of different carriers; dynamically assigned 5QI represents 5G QoS features that need to be sent down to the AN by signaling. 5QI-associated 5G QoS features that describe QoS Flow receives end-to-end message-forwarding processing between the UE and the UPF, which is described in table 3-5.

Table 3-5 5G QoS features

5G QoS features	Descriptions
Resource Type	Includes GBR, delay-sensitive GBR or Non-GBR. It determines whether to allocate dedicated network resources associated with the QoS Flow's guaranteed flow bit rate value. GBR QoS Flows are typically authorized per demand; Non-GBR QoS Flows can be pre-authorized through static policies and billing controls.
Priority Level	Indicates the priority of scheduling resources among QoS Flows. In the case of congestion, when one or more QoS Flows cannot satisfy all QoS requirements, the QoS Flow with a smaller priority value takes precedence over the QoS Flow with a larger priority value. In the absence of congestion, priority should be used to define the resource allocation among QoS Flows. The Dispatcher can prioritize QoS Flow based on other parameters (e.g., resource type, wireless conditions) to optimize application performance and network capacity. Priorities can also be sent to the AN with a standardized or pre-configured 5QI.
Packet Delay Budget, PDB	The upper limit of delay for data packages transmitted between the UE and the anchor UPF. It can be divided into AN PDB and CN PDB, respectively indicating the upper limit of delay on the RAN and Core Network side. The PDB can also be sent to the AN along with the standardized or pre-configured 5QI.
Packet Error Rate, PER	Upper limit for non-congestion related packet loss rate. For delay-sensitive GBR services, packets with delay exceeding the PDB are also counted in the PER.
Average Window	For GBR and delay-sensitive GBR resource types only, used to indicate the duration of GFBR and MFBR calculation at nodes such as UEs, Access Networks and UPFs. Each normalized 5QI (GBR and delay-sensitive GBR resource type) is associated with a default value in the averaging window. The averaging window can also be sent to the AN and UPF along with the standardized or pre-configured 5QI.
Maximum Data Burst Volume, MDBV	Indicates the maximum amount of data that needs to be served over a period of time (AN PDB). Only for latency-sensitive GBR resource types. MDBV can also be sent to AN along with standardized or pre-configured 5QI.

For standardized or pre-configured 5QI, the values of the QoS characteristics corresponding to the 5QI do not need to be transmitted on the interface. The correspondence between standardized 5QI and QoS characteristics can be found in the QoS Model section of 3GPP TS 23.501.

(2) ARP contains information such as priority, preemption capability, and preempted capability. ARP priority defines the relative importance of resource requests. In the case of resource constraint (typically used for admission control of GBR services), ARP is used to determine whether to accept or reject the establishment of a new QoS Flow. It can also be used to determine which existing QoS Flows are preempted during resource constraint. A high priority QoS Flow can preempt the resources of a low priority QoS Flow as capacity allows. An example of ARP functionality is shown in figure 3-35, where QoS Flow A can preempt the resources of QoS Flow C, but not QoS Flow B.

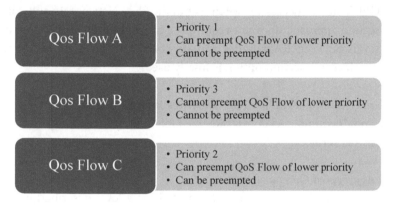

Figure 3-35 Example of ARP Functions

(3) RQA is an optional parameter indicating that reflected QoS is applied to certain traffic carried on this QoS Flow. The SMF can send RQA to the NG-RAN via N2 messages during QoS Flow establishment or modification. See section 3.2.3.3.2 for the reflective QoS mechanism.

(4) GFBR indicates the bit rate that the network is guaranteed to provide for the QoS Flow within the average window. MFBR indicates the maximum bit rate that the QoS Flow can expect to achieve. Exceeded traffic may be dropped or delayed due to rate shaping or policy features of the UE, RAN, and UPF. The network can provide bit rates higher than the GFBR value and not exceeding the MFBR value.

(5) Notification control: Used to indicate whether the NG-RAN needs to notify the Core Network of the event when the GFBR of the QoS Flow is temporarily

not guaranteed (or restored to guarantee again) due to air interface. The details are as follows:

a. For a GBR QoS Flow, if Notification Control is enabled, the NG-RAN sends a notification to the SMF when it believes that the GFBR can no longer be guaranteed and keeps this QoS Flow unreleased. The NG-RAN will attempt to re-guarantee the GFBR unless the NG-RAN requests the release of NG-RAN resources for this GBR QoS Flow in specific conditions (e.g., radio link failure or internal congestion.) The NG-RAN also sends a notification to the SMF when the QoS Flow re-guarantees the GFBR.
b. During the switching process, the target RAN does not know the notification control status of a certain QoS Flow of the source RAN. Two ideas have been discussed in the standard: one is that the source RAN sends the notification control status of the QoS Flow to the target RAN during the switchover; the other one is that the target RAN must be able to guarantee GFBR after a successful switchover of the Core Network default. Since the second approach does not require the transfer of the notification control status between RAN nodes and saves the transfer overhead of the Xn interface, the second option is finally used.

(6) The maximum packet loss rate is the maximum tolerable uplink packet loss rate and downlink packet loss rate in a QoS Flow. It is currently only used for voice mail services.

3.2.3.4.2 Aggregated QoS parameters

In addition to the QoS parameters at the QoS Flow level described earlier, the network also defines QoS parameters at the aggregation level for resource usage control. Their main purpose is to limit the maximum bandwidth that users can use.

The UE Aggregate Maximum Bit Rate (UE-AMBR) limits the aggregated bandwidth of all Non-GBR QoS Flows of a UE and is calculated and controlled by the RAN based on the Session-AMBR, the contracted UE-AMBR, or the UE-AMBR provided by the PCF.

The Session-AMBR limits the aggregated bandwidth of all non-GBR QoS Flows for a specific PDU session. In this case, the UE is responsible for controlling the uplink Session-AMBR; the anchor UPF is responsible for controlling the downlink Session-AMBR and verifying the uplink Session-AMBR; for PDU sessions with multiple anchors, the uplink and downlink Session-AMBRs are aggregated and controlled at multiple anchor UPFs, while the downlink Session-AMBR is also controlled at the UPF anchor point.

3.2.3.5 Summary

The chapter on 5G service quality control explains the basic structure and mechanisms of QoS and the signaling control procedures of QoS Flow and RQC. The SMF executes point-to-point QoS control by sending the PDR and the QER to the UPF, sending QoS configuration data of the QoS Flow to the AN, and sending the QoS Rule to the UE. If RQC is employed, the UE can individually generate the QoS Rule instead of acquiring it from the network side.

5G service quality control also explains the main QoS parameters like 5QI, ARP, and notification controls.

3.2.4 Policy control structure

3.2.4.1 Introduction

The policy control structure is the "nerve center" of the entire communication network architecture and is responsible for making all kinds of complex policy decisions. This nerve center has become increasingly developed as the types of services have become richer. From the early days of billing and QoS control, it has introduced the functions of service flow access technology control and service chain control. In the 5G network, further control of UE policies and access and mobility management policies have been added.

The classification of the 5G network policy control is shown in figure 3-36. It is divided into three categories (access and mobility management policy control, the UE policy control, and session policy control), which are provided by PCF to AMF, UE, and SMF respectively for specific strategy implementation.

3.2.4.2 Policy and charging control structures

The reference framework of the policy and charging control structures for the 5G systems consists of the following functions: PCF, SMF, UPF, AMF, NEF, NWDAF, CHF, AF, and UDR (see figure 3-36).

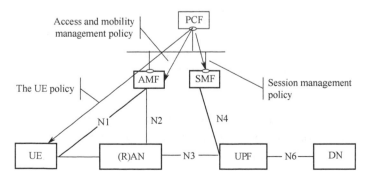

Figure 3-36 Classification of 5G network policy control

Similar to the overall 5G network architecture, a service-based representation and a point-to-point interface-based representation also exist for the 5G policy and charging control structures.

Figures 3-37 and 3-38 show the reference framework of 5G policy and charging control structures for non-roaming scenarios in the form of a servitization-based interface and a point-to-point interface respectively.

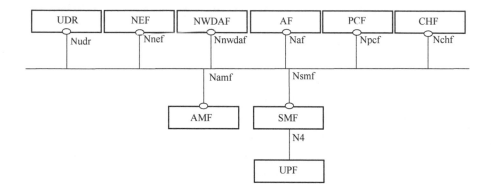

Figure 3-37 5G Policy and charging control structure for non-roaming scenarios (in the form of a servitization-based interface)

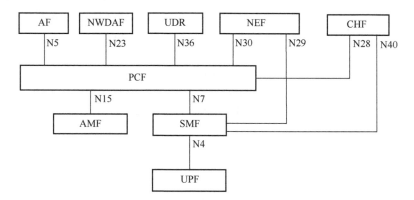

Figure 3-38 5G Policy and charging control structure for non-roaming scenarios (in the form of a point-to-point interface)

Figures 3-39 and 3-40 show the reference framework of 5G policy and charging control structure for roaming LBO scenarios in the form of a service-based interface and a point-to-point interface respectively.

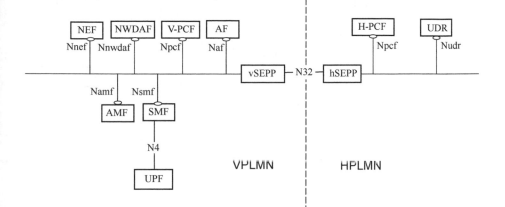

Figure 3-39 5G Policy and charging control structure for roaming LBO scenarios (in the form of a servitization-based interface)

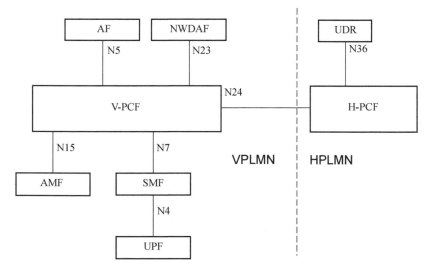

Figure 3-40 5G Policy and charging control structure for roaming LBO scenarios (in the form of a point-to-point interface)

Figures 3-41 and 3-42 show the reference framework of 5G policy and charging control structure for roaming HR scenarios in the form of a service-based interface and a point-to-point interface respectively.

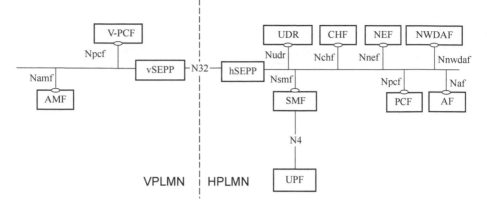

Figure 3-41 5G Policy and charging control structure for roaming HR scenarios (in the form of a servitization-based interface)

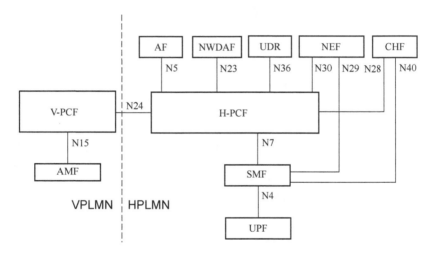

Figure 3-42 5G Policy and charging control structure for roaming HR scenarios (in the form of a point-to-point interface)

3.2.4.3 Basic functions of policy control

3.2.4.3.1 PCF discovery and selection

According to section 3.2.4.2 of this book, AMF and SMF can connect PCFs separately to obtain policies. AMF selects PCFs serving for UE, establishing UE policy association and access mobility management policy association for UE policy control and access and mobility management policy control respectively. SMF selects PCFs for PDU sessions and establishes session management policy association for session management policy control.

1. SELECTING PCF FOR UE

The PCF selected for the UE is used to provide UE policy control and access and mobility management policy control.

The AMF discovers the PCF responsible for UE policy control and access and mobility management policy control based on local configuration or via NRF. In roaming scenarios, the AMF selects H-PCF and V-PCF for policy control.

When the AMF changes, if the new AMF is located in the same PLMN as the old AMF, the old AMF can pass the selected one or more PCF IDs to the new AMF. In a non-roaming scenario, the old AMF sends the PCF ID to the new AMF. If the new AMF cannot establish a connection with the original PCF, the new AMF selects the new PCF to re-establish the policy associated with the PCF. In a roaming scenario, the old AMF sends the V-PCF ID and the H-PCF ID to the new AMF, and if the new AMF cannot establish a connection with the original V-PCF, the new AMF selects the new V-PCF to re-establish the policy associated with the PCF.

2. SELECTING PCF FOR PDU SESSIONS

SMFs provide PCFs with policies for PDU sessions based on local configuration or through NRF discovery. The following factors can be considered when selecting PCFs for SMFs: (1) local operator policies, (2) DNN, and (3) the PCF selected by the AMF (which the SMF can use for session policy control based on the policy).

3. DISCOVERING PCFS FOR AF SESSIONS

The network needs to discover PCFs for AF sessions, whose role is to provide policies for the PDU sessions corresponding to AF sessions. To achieve this function, the Binding Support Function (BSF) is defined in 3GPP TS 23.503 standard.

When there are multiple PCFs deployed in the network, the BSF can ensure that the AF can select the PCF that manages the corresponding PDU session based on the information stored in the BSF. BSF has the following functions:

(1) Storing the information of user identification, the UE address (such as IP address), DNN, and PCF address of PDU session

   a. For IP type PDU sessions, the BSF obtains the corresponding information when an IP address is assigned or released for the PDU session.
   b. For Ethernet-type PDU session, when the MAC address is detected to be used by the UE in the PDU session, the BSF obtains the corresponding MAC address information.

(2) Locating the PCF address based on the request information from AF

(3) Proxying or redirecting the N5 interface request for the UE IP address

3.2.4.3.2 Trigger and execution of policy control

If the policy control structure is considered as the nerve center of the communication system, different network elements in the network will provide and update various input conditions for the decision-making of this nerve center and receive new policies issued by this nerve center.

For access and mobility management policy control, AMF, PCF, and UDR may trigger policy updates. Policy enforcement is performed by the AMF.

For session management policy control, SMF, PCF, UDR, CHF, and AF may trigger policy updates. The policy execution is done by SMF through AN, UPF, and UE.

For UE policy control, AMF, UDR, and PCF may trigger the update of the policy. The execution of the policy is done by the UE.

3.2.4.3.3 Background traffic transmission

The spatial and temporal distribution of traffic in mobile networks is uneven, and there are often more obvious peaks and valleys in the distribution. The purpose of background traffic transmission is to transmit some services with high data volume and insensitive latency during the off-peak time of the network, so as to make full use of the network resources.

The background traffic transmission is initiated by the AF, which can contact any PCF through the NEF and request a time window and some other appropriate conditions for subsequent background traffic transmission.

The AF request will contain information about the amount of data to be transmitted per UE, the desired number of UEs, the desired time window, the external group identification, and optional network area information. In the request notification, the AF may also provide network area information, geographical area, or an area containing a list of NG-RAN nodes and/or a list of cell identifiers. When the AF provides a geographic area, NEF maps it to a list of TAs and/or NG-RAN nodes and/or cell identifiers based on local configuration.

The PCF is responsible for developing a background transmission policy. The policy consists of a recommended background traffic transmission time window, a rate reference for that time window, a background traffic transmission reference identifier, and network area information. Finally, the PCF returns the candidate list of background transmission policies or the selected background transmission policy to the AF through the NEF. If the AF receives more than one background transmission policy, the AF selects one of them and informs the PCF of the selected background transmission policy.

The background traffic transmission policy negotiation flow is shown in figure 3-43.

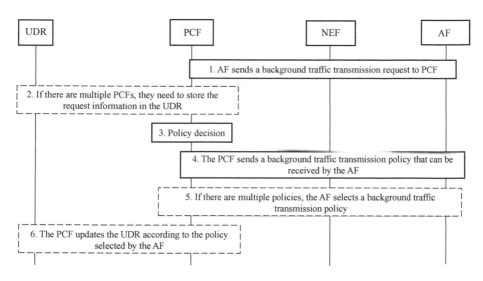

Figure 3-43 Background traffic transmission policy negotiation flow

When the AF wants to apply the background traffic transfer policy to an established PDU session, the AF will provide the background traffic transfer reference mark to the PCF along with the AF session information when the background traffic transfer is about to start. The PCF obtains the corresponding background transfer policy from the UDR and determines the corresponding PCC rules.

When the AF wants to apply the background traffic transport policy to future PDU sessions, the AF provides the background traffic transport reference ID to the NEF with the external identifier or external group identifier of the affected UE, and the NEF stores the background traffic transport reference ID in the UDR as the application data of the UE. The PCF serving the UE can obtain the application data of the UE as input through the UDR and send the relevant background traffic transmission information to the UE through the time window and location interval in the UE routing policy, so that the corresponding background traffic transmission can be initiated by the UE in the future.

3.2.4.4 UE policies

3.2.4.4.1 Types of UE policy

UE policy information includes the following two categories, which can be found in sections 3.2.4.4.2 and 3.2.4.4.3 of this book:

(1) UE Route Selection Policy (URSP): used by the UE to determine whether the application can be associated with an established PDU session or needs to trigger the establishment of a new PDU session. The URSP can be provided by the PCF in the HPLMN or can be pre-configured on the UE. When both the pre-configured URSP and the URSP provided by the PCF exist on the UE, the UE uses only the URSP provided by the PCF.

(2) Access Network Discovery & Selection Policy (ANDSP): used for UE to select WLAN Access Network, ANDSP can be provided by PCF. In the roaming scenario, the V-PCF and H-PCF can provide their respective ANDSPs; when the UE is located in HPLMN, the ANDSP provided by the PCF of HPLMN is used; in the roaming scenario, when the UE contains the ANDSPs provided by the VPLMN PCF and HPLMN PCF, the UE gives priority to the ANDSP provided by the PCF of VPLMN.

3.2.4.4.2 UE Route Selection Policy (URSP)

The UE Route Selection Policy (URSP) consists of at least one URSP rule. When valid USRP rules exist in the UE, the UE matches applications to these rules to determine how to route uplink data packets.

Each URSP rule contains a rule priority value that determines the priority of the rule in the policy.

Each URSP rule contains a service descriptor that determines when the rule should apply. Service descriptors include the following types:

(1) Application descriptor: contains the operating system identifier and the application identifier on that operating system

(2) IP descriptor: the destination IP triplet (destination IPv4 address or IPv6 network prefix, destination port number, and protocol ID)

(3) Non-IP descriptor: the destination information of non-IP services

(4) Domain Descriptor: the destination Fully Qualified Domain Name (FQDN)

(5) DNN: the DNN information provided by the application

(6) Connection capability: the information provided by the UE application when requesting a network connection with a specific capability

Each URSP rule contains a list of route selection descriptors for one or more routing descriptors, each with a different route selection descriptor priority value. A route selection descriptor contains one or more of the following components:

(1) Session and Service Continuity mode: indicates that matching services should be routed through PDU sessions that support the session and service continuity mode.

(2) Network Slice Selection: indicates that the matching service will be routed through a PDU session that supports any S-NSSAI that includes one or more S-NSSAIs.

(3) DNN Selection: indicates that the matched services will be routed through a PDU session that supports any of the contained DNNs, which includes one or more DNNs.

(4) Non-Seamless Diversion Indication: indicates that the matched service will be diverted to a non-3GPP access outside the PDU session when the rule is applied. If this component is present in the routing descriptor, no other components are included in the route selection descriptor.

(5) Access Type Preference: indicates the preferred access type (3GPP, non-3GPP, etc.) when establishing a PDU session.

When PCF issues a URSP rule to the UE, it can include a URSP rule with a wildcard service descriptor.

The URSP rule with a wildcard service descriptor is used to route services that do not match any other URSP rule application. Therefore, this should be used as the last URSP rule, in other words, the one with the lowest priority; otherwise URSP rules, with a lower priority, will not be matched.

The URSPs in the UE are shown in figure 3-44, where different S-NSSAIs are indicated by different types of shading.

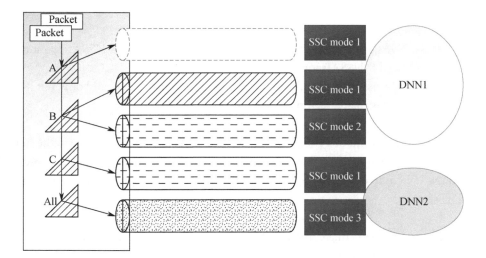

Figure 3-44 URSP in the UE

The logic by which the UE executes the URSP is shown in figure 3-45.

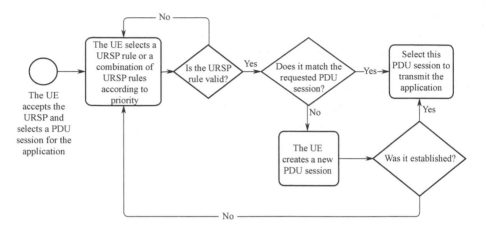

Figure 3-45 The logic by which the UE executes the URSP

For each newly detected application, the UE evaluates the URSP rule in the order of rule priority and determines whether the application matches the service descriptor of any URSP rule.

When the application matches the service descriptor in the URSP rule, the UE will select the route selection descriptor in that URSP rule in the order of the route selection descriptor, and the selected descriptor will only be valid if the usage conditions are met, such as:

(1) an S-NSSAI can only be considered valid if it exists and the S-NSSAI is among the allowed S-NSSAIs; or
(2) if a DNN exists and is a LADN DNN, the UE is considered valid only if it lies within the available region of that LADN.

When a valid route selection descriptor is found, the UE determines whether an existing PDU session exists that matches all components in the selected descriptor, and the UE compares the components in the selected descriptor with the existing PDU session:

(1) For components that contain only one value (such as SSC mode), the value of the PDU session must be the same as the value specified in the route selection descriptor.
(2) For components that contain a list (such as network slice selection), the value of the PDU session must belong to the value specified in the route selection descriptor.

When a matching PDU session exists, such as Service C in figure 3-44, the UE associates the application to an existing PDU session. In other words, it routes the detected services of the application to that PDU session.

If the UE determines that multiple matching PDU sessions exist, such as Service B in figure 3-44 with only one DNN specified, and multiple existing PDU sessions for different network slices all match the requirements, the UE may select one of them for use based on the implementation.

If all PDU sessions do not match, the UE tries to establish a new PDU session using the value specified by the selected routing descriptor, e.g., if the PDU session corresponding to Application A in figure 3-44 does not exist, then UE will initiate a PDU session to establish the process; if the PDU session establishment request is accepted, the UE associates the application to this new PDU session; if the PDU session establishment request is rejected, depending on the rejection reason, the UE selects another value in the selected routing descriptor for combination, or selects the next route selection descriptor in the order of route selection descriptor priority.

3.2.4.4.3 Access Network discovery and selection policy (ANDSP)
In the first phase of the 5G standard, the Access Network discovery and selection policy only contains rules to assist the UE in selecting a WLAN Access Network. It does not support selection rules for other types of non-3GPP Access Networks.

The WLAN Access Network selected by the UE using the Access Network discovery and selection policy can be used for direct streaming (i.e., sending traffic to the WLAN outside the PDU session) and registered to the 5GC using the non-3GPP Access Network selection information.

If the UE supports non-3GPP access to the 5GC, it needs to support the ANDSP.

The Access Network discovery and selection policy will contain one or more WLAN Selection Policy (WLANSP) rules and may also contain information about the UE's selection of the Non-3GPP Interworking Function (N3IWF).

In both non-roaming and roaming scenarios, the UE can obtain the ANDSP from the PCF in the HPLMN. In roaming scenarios, the PCF in the VPLMN can also provide the ANDSP. Since the UE in roaming scenarios accesses from the VPLMN, it is generally assumed that the VPLMN is more aware of its WLAN Access Network deployment than the HPLMN. Therefore, the ANDSP provided by the VPLMN has higher priority than that provided by the HPLMN.

3.2.4.5 Access and mobility management policy
The access and mobility management policy includes service area restriction management and wireless resource management. This part will be stored in the AMF as a UE context.

In non-roaming scenarios, the UE can obtain the access and mobility management policy from the PCF of the HPLMN. In roaming scenarios, the access and mobility management policy is provided by the PCF in the VPLMN.

Service area restriction management means that the PCF can modify the service area restriction used by the AMF. The service area restrictions include the list of allowed TAIs or the list of disallowed TAIs, and the optional maximum number of TAIs. The PCF can instruct the AMF to have unlimited service areas. The PCF can modify the pre-configured policies based on the information provided by the AMF (e.g., UE location or signed service area restrictions) by expanding the list of allowed TAIs, reducing the number of disallowed TAIs, or increasing the maximum number of allowed TAIs. The AMF receives and stores the modified service area limit from PCF and then uses it to determine the UE's mobility limits. In addition, the PCF can update the service area limits based on information provided by other NFs.

Wireless resource management means that the PCF can modify the Radio Access Technology/Frequency Selection Priority (RAT/RFSP) index. The PCF can modify the RFSP index based on operator policies (e.g., cumulative usage, or load level information for each network slice instance), or the signed RFSP index provided by the AMF. The AMF sends the PCF's updated RFSP index to the RAN.

3.2.4.6 Session management policy

3.2.4.6.1 Session management policy information

Session management-related policies are classified into charging, QoS control, data forwarding, and other related aspects, which are controlled by the PCF associated with the SMF. In the roaming LBO scenario, the session management-related policy control is handled by the V-PCF associated with the V-SMF. The PCF in the VPLMN can interact with the AF to generate PCC rules for services issued through the VPLMN. The PCF in the VPLMN does not access the user policy information of the HPLMN but uses the locally configured policy as PCC rule generation input. The AF in the LBO scenario in the attributed PLMN is not supported in 5G for the time being because the commercial scenario is not clear. For the HR scenario, the policy control related to session management is handled by the H-PCF connected to the H-SMF.

The PCF can send PCC rules and PDU session-level policy information to the SMF to achieve appropriate session management policy control.

PCC rules are the information required to enable user-plane detection, policy control, and reasonable charging for service data flow. PCC rules contain the following contents:

(1) Rule identification: identifies one PCC rule within a PDU session. Used for PCC rule reference between the PCF and SMF.

(2) Service data flow detection information: defines the message detection method of service data flow, such as the service data flow template.

(3) Charging information: information such as charging key, sponsor identification, and charging method.

(4) QoS control and gating information: indicates how to apply QoS control and gating of service data streams, which includes gate status, maximum bit rate, and redirection.

(5) Access Network reporting information: describes the Access Network information reported by PCC rules when QoS Flow is established, modified, or deleted, such as user location and user time zone.

(6) Usage monitoring control information: refers to the identification required when using usage monitoring, such as Monitoring Key (monitoring key value), and session-level monitoring mutual exclusion indication.

(7) Service chain information: describes the information required to perform service chain control.

(8) RAN support information: information that supports RAN decisions (such as switching threshold judgments), such as the maximum packet loss rate.

The purpose of the PDU session-related policy information is to provide information related to policy and charging control applicable to a single Monitoring Key or an entire PDU session. The contents contained in the PDU session-related policy information are as follows:

(1) Charging information: contains the CHF address

(2) Default charging method: the default charging method of the PDU session

(3) Policy control request trigger: defines the event that causes the PCC rule to be re-requested for the PDU session

(4) Re-verification time limit: the minimum time period between PCC rule requests executed by the SMF

(5) Presence Reporting Area (PRA)-related information: defines the reporting area

(6) IP index: provided to the SMF to assist in determining the IP address assignment method for PDU sessions

(7) PDU session-level QoS information: such as explicitly indicated QoS features, reflective QoS timers, and authorized Session-AMBR

(8) Usage monitoring control-related information: defines the information required for usage monitoring, such as monitoring time and traffic threshold

3.2.4.6.2 Binding mechanism

The binding mechanism is the basic mechanism for session management policy control, and its purpose is to associate a service data flow with a QoS Flow. The service data flow is identified by the service data flow template in the PCC rule.

The specific binding mechanism consists of the following three steps:

1. SESSION BINDING

Session binding is the association of an AF session with a PDU session. Session binding is applicable to both IP-type and Ethernet-type PDU sessions. The following parameters can be considered by PCF for session binding: (1) UE IPv4 address, IPv6 network prefix, or UE MAC address; (2) UE identity, e.g., Subscription Permanent Identifier (SUPI); (3) data network (DN) information, i.e., DNN.

After identifying the affected PDU sessions, the PCF further identifies the PCC rules affected by AF session information, including new PCC rules to be created and existing PCC rules that need to be modified or deleted.

2. PCC RULE AUTHORIZATION

PCC rule authorization is the selection of 5G QoS parameters for PCC rules.

For PCC rule authorization, PCF will consider any 5GC-specific restrictions, signing information, and other information that can be used as input. Each PCC rule will receive a set of QoS parameters supported by the specific Access Network. For emergency services, the PCF can authorize without signing information.

3. QoS FLOW BINDING

QoS Flow binding refers to a PCC rule associated with a QoS Flow within a PDU session, with the following binding parameters: (1) 5QI, (2) ARP, (3) notification control, (4) priority, (5) average window, (6) MDBV.

When the PCF issues a PCC rule, if the PCF requests to bind the PCC rule to the QoS Flow associated with the default QoS Rule, then the SMF binds the PCC rule directly to the QoS Flow associated with the default QoS Rule. In other cases, the SMF will evaluate whether a QoS Flow exists with the same binding parameters; if not, the SMF uses the parameters in the PCC Rule to deduce the QoS parameters and generate a new QoS Flow to bind the PCC Rule to the QoS Flow; if a QoS Flow exists with the same binding parameters, then the SMF updates the QoS Flow so that the new PCC Rule is associated with that QoS Flow. The SMF must re-evaluate the existing binding whenever the authorized QoS parameters of a PCC rule change.

New 5G parameters include notification control, priority, average window, and maximum data burst volume. Priority, average window, and max data burst volume are features in 5QI and can be modified by the PCF. 5QI is one of the binding parameters, so they will also become binding parameters. Notification control, which indicates whether GFBR requests notification from 3GPP when it cannot (or again) satisfy QoS Flow during the QoS Flow lifecycle, i.e., notification control is a basic QoS Flow feature, so QoS Flow binding also needs to consider notification control.

3.2.4.7 Summary

Policy control includes the basic framework and some basic functions of 5G policy control, including the discovery and selection of the PCF, and the triggering and execution of policy control.

Policy control also includes the structure and application of major policy rules, such as the UE policy, the access and mobility management policy, and the session management policy. The UE policy carries out routing, Access Network discovery, and selection control; the access and mobility management policy carries out service area restriction management and wireless resource management control; the session management policy carries out charging, data forwarding, and QoS control.

3.2.5 Voice

3.2.5.1 Introduction

While data services have driven the evolution of 5G, voice services remain an important service for operators. As with the deployment of the 4G networks, 5G commercial deployments need first to determine how voice services will be delivered. In addition, the NG-RAN includes NR, evolved LTE, and a combination of both. For ease of description, this section uniformly uses Voice over NR (VoNR) to refer to voice solutions that carry voice on the NG-RAN.

Considering that the 4G LTE-based voice service (Voice over LTE, VoLTE)/IP Multimedia Subsystem (IP Multimedia Subsystem, IMS) has already been commercially deployed, the 5G voice solution needs to focus on how to evolve based on the deployed 4G VoLTE/IMS network in order to achieve a smooth evolution of voice services. In the long run, VoNR is the ultimate solution for 5G voice services. However, individual operators need to choose the appropriate voice solution at different stages of development based on the characteristics of their own networks:

(1) EPS Fallback solution (5G does not support VoNR, 4G supports VoLTE): considering that the early deployment of 5G cannot support VoNR temporarily, similar to the 4G voice solution, it is necessary to consider how to fall back to the deployed 4G VoLTE and provide 5G voice service quickly, i.e., the EPS Fallback solution.

(2) VoNR solution (5G supports VoNR, 4G supports VoLTE): Considering that 5G coverage takes some time to reach or exceed 4G coverage, the interoperability of VoNR and VoLTE needs to be supported to ensure the continuity of voice service. It only needs to replace the eNB and EPC of the 4G network with the NG-RAN and 5GC of the 5G network.

(3) VoNR → 3G Single Radio Voice Call Continuity (SRVCC) solution (5G does not support VoNR, 4G does not support VoLTE or 4G is not deployed): Considering that a small number of operators who do not deploy 4G VoLTE or do not deploy 4G wish to provide voice continuity through the 3G CS Domain, it may be necessary to consider interoperability of VoNR and 3G network CS Domain, i.e., a 5G → 3G SRVCC solution.

If the 5G network supports VoNR or EPS Fallback, the UE goes through the network registration process at IMS through the 5G network, otherwise the UE will register in the 4G or 2G/3G network.

Next, the voice scenarios for the 5G systems are explained separately according to specific network deployment scenarios.

3.2.5.2 EPS Fallback solution

The EPS Fallback solution is shown in figure 3-46. Early 5G deployment cannot support VoNR for the time being, and it is necessary to consider how to fall back to the deployed 4G VoLTE to provide 5G voice services quickly.

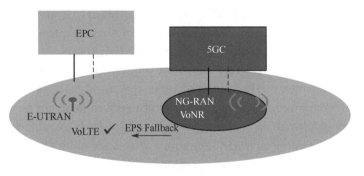

Figure 3-46 EPS Fallback

The EPS Fallback flow chart is shown in figure 3-47. The UE first initiates the VoNR call flow in 5G, then the NG-RAN rejects the received voice bearer establishment request and moves the UE to the 4G network to fall the voice service back to 4G VoLTE.

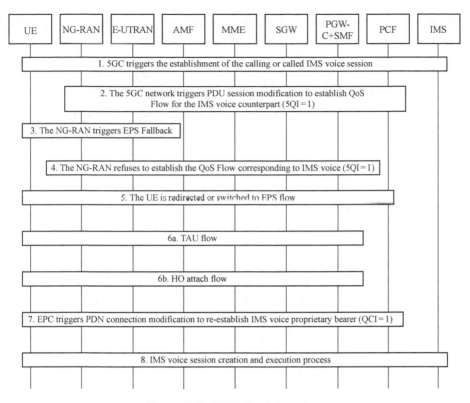

Figure 3-47 EPS Fallback flow chart

Step 1: The UE initiates IMS-based calling or called VoNR service in the 5G network.

Step 2: The IMS network interacts with the 5GC to trigger the establishment of a QoS Flow (5QI = 1). The NG-RAN receives the QoS Flow (5QI = 1) establishment request.

Step 3: If the NG-RAN supports EPS Fallback, the NG-RAN triggers the EPS Fallback process, i.e., UE fallback to the EPS process based on the UE capability information, network configuration (e.g., the network supports N26 interface), the "Support Redirect EPS Fallback" indication from the AMF, and wireless conditions.

Step 4: The NG-RAN does not support VoNR, so it rejects the QoS Flow (5QI = 1) establishment request.

Step 5: NG-RAN falls the UE back to the 4G network by 5G to 4G PS switching or redirection.

Step 6a: If the network supports the N26 interface, the UE performs the TAU process after the network has backed off the UE to the 4G network via 5G to 4G PS switchover or redirection.

Step 6b: If the network does not support the N26 interface, the UE initiates an HO attach process in the 4G network to keep the UE's IP address unchanged, ensuring a successful IMS voice call after the network has backed off the UE to the 4G network via a redirection process (see 3GPP TS 23.401 [6], section 5.3.2.1 for details).

Step 7: After the UE has fallen back from the 5G network to the 4G network, the PGW-C+SMF re-initiates the IMS voice proprietary bearer (QCI = 1) establishment process.

Step 8: The IMS voice call is successfully established at the EPS.

3.2.5.3 VoNR Solution

5G coverage takes some time to reach or exceed 4G coverage, so interoperability of VoNR and VoLTE needs to be supported to ensure the continuity of voice services.

The VoNR 5G to 4G PS HO is shown in figure 3-48, where the UE initiates a VoNR call in the 5G network and the NG-RAN receives a QoS Flow (5QI = 1) establishment request. The NG-RAN already supports VoNR, so the NG-RAN allocates relevant airspace resources for voice services and carries the voice services in the 5G network. If the UE moves out of the 5G coverage area, the NG-RAN will trigger the 5G → 4G PS switchover to switch the voice service to 4G VoLTE to ensure the continuity of voice service.

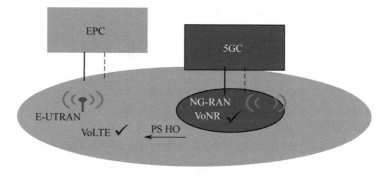

Figure 3-48 VoNR 5G to 4G PS HO

3.2.5.4 VoNR → 3G SRVCC Solution

A small number of operators who do not deploy 4G VoLTE or 4G want to provide voice continuity service through 3G network CS Domain, so they also need to consider the interoperability of VoNR and 3G network CS Domain. Currently, only 5G → 3G SRVCC is supported, and 5G → 2G SRVCC is not supported for the time being.

The 5G to 3G SRVCC is shown in figure 3-49. The UE first initiates a VoNR call in the 5G network, and the NG-RAN allocates relevant airspace resources for voice services and carries the voice services in the 5G network. If the UE moves out of 5G coverage area, NG-RAN will trigger 5G → 3G SRVCC switching the 5G voice service to 3G CS because 4G does not support VoLTE or 4G is not deployed.

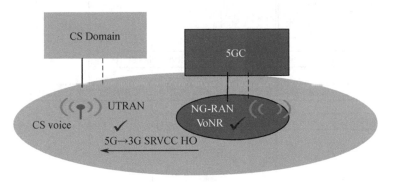

Figure 3-49 5G to 3G SRVCC

The network architecture of 5G to 3G SRVCC is shown in figure 3-50. Since 5G does not support direct interoperability with 3G, in order to maximize the reuse of the existing 4G SRVCC-related architecture, 5G → 3G SRVCC is implemented based on the following indirect architecture deployment: the N26 interface is supported between AMF and MME_SRVCC, and the Sv interface is supported between MME_SRVCC and MSC.

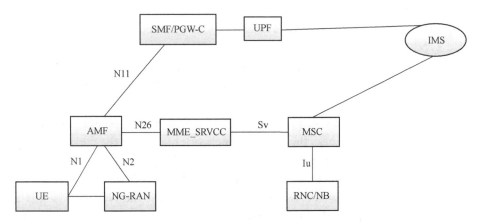

Figure 3-50 Network architecture of 5G to 3G SRVCC

The process of 5G to 3 G SRVCC is shown in figure 3-51.

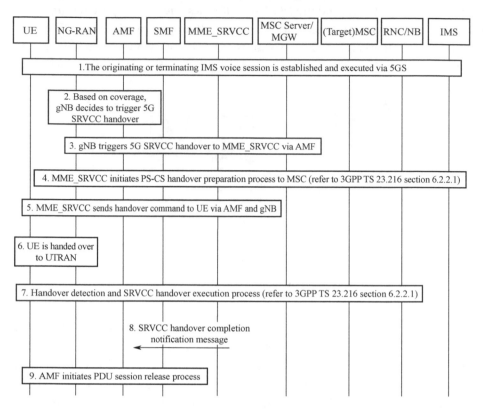

Figure 3-51 Process of 5G to 3G SRVCC

Step 1: The UE initiates the VoNR call flow.

Step 2: The NG-RAN triggers 5G → 3G SRVCC switchover based on reasons such as NG-RAN coverage.

Step 3: The NG-RAN sends an SRVCC switch request to the AMF, and the AMF sends a forward relocation request message to the MME_SRVCC to forward the switch request.

Step 4: The MME_SRVCC triggers the PS to CS switchover preparation process to the MSC, which asks the target 3G network to prepare the corresponding radio resources for the voice call. The details of this process can be seen in the 4G SRVCC switching process in section 6.2.2.1 of 3GPP TS 23.216 [7].

Step 5: When the target-side radio resources are ready, the NG-RAN sends a switch command to the UE, instructing the UE to switch to the 3G network.

Step 6: The UE accesses the UTRAN network.

Step 7: The target 3G network detects UE access, and the CS domain Core Network executes PS to CS switching flow, the details of which can be found in the 4G SRVCC switching process defined in section 6.2.2.1 of 3GPP TS 23.216 [7].

Step 8: After the PS to CS switch is executed, MME_SRVCC notifies the AMF that the SRVCC switch has been completed.

Step 9: The AMF triggers the deletion of all PDU sessions associated with the UE in 5GC.

3.2.5.5 Summary

5G commercial deployment needs first to determine how to provide voice services. At the same time, in order to achieve smooth evolution of voice services, 5G voice solutions need to focus on how to evolve on top of deployed 4G VoLTE/IMS networks. In the long run, VoNR is the ultimate solution for voice services in 5G networks, but operators need to choose the appropriate 5G voice solution according to their own network conditions at various development stages, i.e., the EPS Fallback solution, VoNR solution, or 5G → 3G SRVCC solution.

3.2.6 Combination networking and interoperability

3.2.6.1 Introduction

Since 5G network coverage is not perfect at the early stage of the 5G network deployment, 4G network coverage is needed to supplement data service coverage to ensure the continuity of user services. Therefore, 5G combination networking and interoperability is one of the key considerations for operators when deploying 5G networks.

5G combination networking and interoperability includes 5G combined networking and interoperability architecture, interoperability process in scenarios with and without an N26 interface. In addition, 5G combination networking and interoperability is also the basis for the 5G voice scenarios such as EPS Fallback described in section 3.2.5 of this book.

3.2.6.2 Combination networking and interoperable architecture for 5G

A non-roaming architecture supporting interoperability between 5G and 4G is shown in figure 3-52.

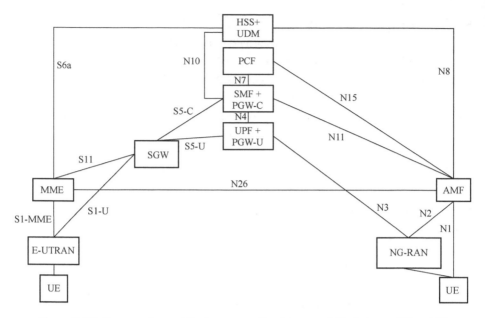

Figure 3-52 Non-roaming architecture supporting interoperability between 5G and 4G

Figure 3-52 shows that

(1) the N26 interface between the AMF and the MME is optional;
(2) HSS+UDM, SMF+PGW-C and UPF+PGW-U collocated network elements
     are used for interoperability between 5G and 4G depending on the capability
     or contracting of the UE; and if the UE does not support interoperability,
     these collocated network elements are not required to serve it; and
(3) an additional UPF may exist between the NG-RAN and the UPF+PGW-U.
     For example, it may use an Uplink Classifier (UL CL) for splitting or relaying
     data traffic between NG-RAN and UPF+PGW-U.

The UE needs to support both 5G and 4G NAS in order to support 5G
interoperability with 4G. Also, during 5G network registration, the 5G UE needs
to instruct the network UE to support both 5G and 4G NAS. The 5G UE supports
both Single Registration (SR) and Dual Registration (DR) modes. In SR mode,
the UE has only one active MM state (RM-5GC, EMM-EPC); the UE can only
register in 5G or 4G networks, not in both. Second, when in DR mode, the UE can
be registered in the 5G network alone, the 4G network alone, or both 5G and 4G
networks.

3.2.6.3 Interoperability in scenarios with an N26 interface

An N26 interface is used to provide seamless session continuity for single registration mode UEs, e.g., for voice services. In single registration mode, the IP address of the 5G and 4G interoperability process remains unchanged based on the premise assumption of SSC mode 1.

3.2.6.3.1 Switching flow between 5G and 4G

In the connected state, when the UE moves from the 5G network to the 4G network, it performs the 5G to 4G switching flow, as shown in figure 3-53. See section 4.11.1.2.1 of 3GPP TS 23.502 [2] for details.

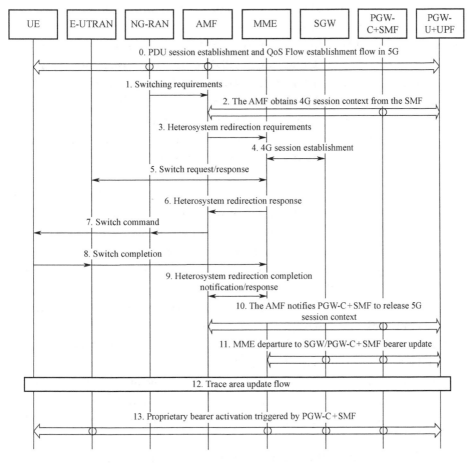

Figure 3-53 The UE moves from the 5G network to 4G network, and the UE switches between 5G and 4G

Step 0: During PDU session establishment and GBR QoS Flow establishment in the 5G network, if the PCF is deployed, PGW-C+SMF performs 4G QoS mapping from the 5G QoS parameters obtained from the PCF, and assigns the PCC rules obtained from PCF to TFT; otherwise, 4G QoS mapping and TFT assignment are mapped locally by PGW-C+SMF, and PGW_C+SMF ignores 5G QoS parameters that are not suitable for 4G (e.g., QoS notification control).

   If the SMF determines that an EPS Bearer ID needs to be assigned to the QoS Flow, the SMF requests the service's AMF to assign it, as described in section 4.11.1.4 of 3GPP TS 23.502 [2].

Step 1: The 5G base station decides to switch to the 4G base station based on information such as the measured signal quality of the UE and initiates a switch request to the AMF.

Step 2: The AMF obtains from the PGW-C+SMF the 4G session context information generated by the UE during the PDU session establishment process.

Step 3: The AMF sends a heterosystem redirection request message to the MME, carrying the 4G mobility context information generated by the AMF and the 4G session context information obtained from PGW-C+SMF.

Step 4: The MME establishes a 4G session based on the 4G session context information, by obtaining tunnel information from the SGW for each bearer on the SGW in the session.

Step 5: The MME sends a switch request to the E-UTRAN, and the E-UTRAN gives feedback to the MME on the bearer information and the tunnel information on the E-UTRAN for which the air interface agrees to switch.

Step 6: The MME sends a heterosystem redirection response message to the AMF.

Step 7: The AMF sends the switch command to the UE via the NG-RAN.

Step 8: The UE notifies the MME via E-UTRAN that the switchover is complete.

Step 9: The MME notifies the AMF that the heterosystem redirection is successful, and the AMF needs to respond to this message.

Step 10: The AMF notifies the PGW-C+SMF to keep the 5G side session context and release the tunnel on PGW-U+ UPF.

Step 11: The MME triggers a bearer update between the SGW as well as PGW-C+SMF based on the bearer information from E-UTRAN that agrees to the switch in step 5, i.e., establishes the bearer between E-UTRAN and SGW/PGW-C+SMF.

Step 12: The UE initiates the tracking area update process.

Step 13: For other proprietary bearers not established in the above steps, PGW-C+SMF initiates the bearer establishment process.

If the UE moves from 4G to 5G, it performs the 4G to 5G switchover flow as shown in figure 3-54. See section 4.11.1.2.2 of 3GPP TS 23.502 [2].

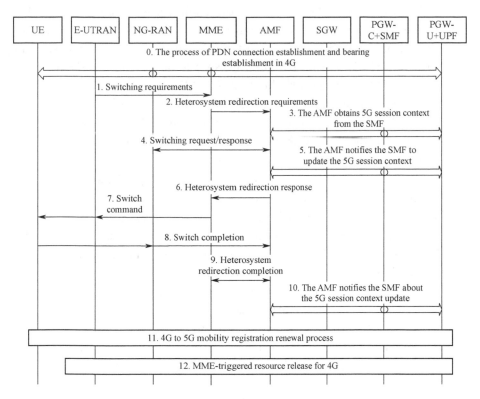

Figure 3-54 The UE moves from 4G to 5G, and switches from 4G to 5G

Step 0: When the UE is served by the 4G network, during the PDN connection establishment process, the UE allocates a PDU Session ID, which is sent to the PGW-C+SMF through the Protocol Configuration Option (PCO). If the PCC is deployed, during the PDN connection establishment and proprietary bearer establishment process, the PGW-C+SMF will perform 4G QoS mapping from the 5G QoS parameters obtained from the PCF and obtain TFTs from PCF. Otherwise, 4G QoS and TFTs are generated locally by the PGW-C+SMF. Other 5G QoS parameters corresponding to this PDN connection, such as Session-AMBR, QoS Rules, and QoS parameters at QoS Flow level, can be sent to the UE in PCO. 5G QoS parameters are stored in the UE for use when the UE switches from 4G to 5G.

Step 1: The 4G base station decides to switch to the 5G base station based on information such as the measured signal quality of the UE, and initiates a switch request to the MME.

Step 2: The MME sends a heterosystem redirection request message to the AMF.

Step 3: The AMF obtains the 5G session context information generated by the UE during the PDN connection establishment process and the session information (PDU Session ID and S-NSSAI) to be stored on AMF from the PGW-C+SMF, during which the PGW-C+SMF notifies the PGW-U+UPF to establish the tunnel information and sends it to SMF together.

Step 4: The AMF sends a switch request to the NG-RAN, and the NG-RAN responds to the AMF with the session information (including QoS Flow information and tunnel information on the NG-RAN), agreeing to the switch.

Step 5: The AMF sends the session information (including QoS Flow information and tunnel information on the NG-RAN) that the NG-RAN agrees to switch to the PGW-C+SMF and PGW-U+UPF.

Step 6: The AMF sends a heterosystem redirection response message to the MME.

Step 7: The MME sends the switch command to the UE via E-UTRAN.

Step 8: The UE notifies the AMF via the NG-RAN that the switchover is complete.

Step 9: The AMF notifies the MME that the heterosystem redirection is successful. MME needs to respond to this message.

Step 10: The AMF notifies the PGW-C+SMF to update the 5G session context information, and since the switchover is already successful at this point, this step mainly instructs the PGW-C+SMF to release the temporary data forwarding tunnel.

Step 11: The UE initiates 4G to 5G mobility registration renewal process requests.

Step 12: The MME triggers the PDN connection on the 4G side and the bearer release.

3.2.6.3.2 Idle state mobility process between 5G and 4G

In the idle state, if the UE moves from 5G to 4G, it performs the TAU process at 4G. With the N26 interface, the UE mobility flow from 5G to 4G in the idle state is shown in figure 3-55. See section 4.11.1.3.2 of 3GPP TS 23.502 [2].

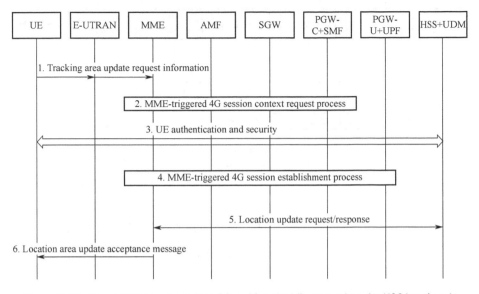

Figure 3-55 Flow of UE movement from 5G to 4G in the idle state when the N26 interface is available

Step 1: The UE sends a tracking area update request message to the MME via E-UTRAN, which indicates that the MME and the UE have moved from 5G to 4G.

Step 2: The MME requests the 4G session context from the AMF, which further obtains the 4G session context generated by 5G during the PDU session establishment from the PGW-C+SMF, and sends it to the MME.

Step 3: The UE authentication and security process.

Step 4: The MME establishes the PDN connection and the corresponding bearer on the 4G side based on the 4G session context from the AMF.

Step 5: The MME sends a location update request to the HSS+UDM, which responds to the request.

Step 6: The MME sends a tracking area update acceptance message to the UE.

If the UE moves from 4G to 5G, it performs the Mobility Registration process at 5G. The mobility flow of the UE from 4G to 5G in the idle state when the N26 interface is available is shown in figure 3-56. See section 4.11.1.3.3 of 3GPP TS 23.502 [2].

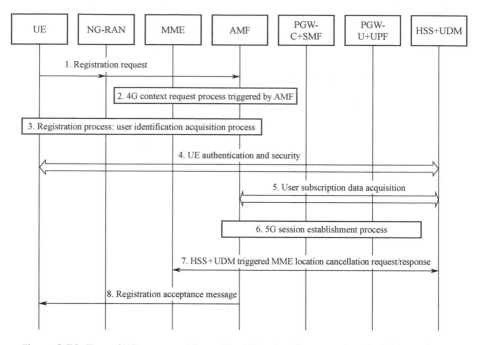

Figure 3-56 Flow of UE movement from 4G to 5G in the idle state when the N26 interface is available

Step 1: The UE sends a registration request message to the AMF via the NG-RAN, which indicates that the AMF and the UE have moved from 4G to 5G.

Step 2: The AMF requests the 4G session context and mobility management context from MME, and AMF locally maps the 4G mobility management context to the 5G mobility management context.

Step 3: If the AMF does not obtain a UE identity in step 2, it needs to obtain a UE identity, e.g., SUCI, from the UE through the UE identity acquisition step.

Step 4: The UE authentication and security process.

Step 5: The AMF obtains the UE sign-up data from the HSS+UDM based on the user identification.

Step 6: The AMF initiates the 5G session establishment process to the PGW-C+SMF based on the 4G session context from step 2.

Step 7: The HSS+UDM triggers a location cancellation request/response from the MME.

Step 8: The AMF sends a registration acceptance message to the UE.

3.2.6.4 Interoperability in scenarios without the N26 interface

In the absence of the N26 interface, the UE provides IP address reservation for inter-system mobility by storing and acquiring PGW-C+SMF and corresponding APN/DNN information via the HSS+UDM.

The UE can function in single registration mode or dual registration mode. If the N26 interface is not supported in the network, the network side needs to instruct the UE network side to support DR during the UE registration process, and the UE can register in the destination system in advance after getting this instruction information.

3.2.6.5 Summary

5G combined networking and interoperability schemes is an important means to ensure 5G service continuity and can be generally divided into two major scenarios according to the existence of N26 interfaces between AMF network elements and MME network elements: (1) in the case of N26 interface support, the UE performs only single registration; or (2) in the case that the N26 interface is not supported, the UE can perform either single registration or dual registration.

In 5G Rel-15 and Rel-16 phases, 5G only supports interoperability with 4G, and not with 2G/3G.

3.3 Service-based architecture

3.3.1 Introduction

The 5G white paper published by the Next Generation Mobile Networks (NGMN) organization [8] has many requirements for the deployment, operation, and management of the 5G networks. One is that 5G networks need to provide the ability to introduce new services and technologies efficiently, flexibly, and quickly to accommodate future market demands.

The introduction of a new service typically requires the network to support new capabilities. Operators build this new capability into their networks, either by deploying new equipment or by integrating existing network capabilities to create new capabilities, the latter of which is clearly more economical for operators. Therefore, whether the definition and granularity of network functions are flexible enough to enable new network functions by integrating existing functions flexibly is an important concern in the design of the 5G network architecture.

The 5G system provides users with access services to various services. From a network perspective, access services include attributes such as mobility, security authentication, and connection reliability, and each access attribute corresponds to a functional module of the 5G system. For example, keeping the service uninterrupted while the user is on the move relies on the network's mobility management function and session management function.

To achieve a flexible and scalable 5G architecture, the overall functionality of the 5G system needs to be broken down into functional modules and be deployed at the granularity of such modules. This has been the consensus of many operators since the beginning of 5G discussions.

Looking back at the Core Network (EPC) architecture of the 4G networks, the complex and overlapping functional combinations of network elements make it impossible to customize the control function combinations for a specific service type. As a result, all services will share the same set of logical control functions. The tight coupling between many control functions and the complexity of interfaces among network elements makes it extremely difficult to bring services online and operate and maintain the network, which is not flexible enough to support the multi-service scenarios of the 5G era.

During early discussions about 5G Core Network architecture, a conception was mentioned frequently, namely Cloud Native—a new system practice paradigm for building, running, and managing Cloud-based software in a Cloud environment by leveraging Cloud infrastructure and platform services. It has already been used in many commercial cases in the IT industry. Cloud Native-based telecom Cloud networks have several key technologies: stateless design, microservice decoupling, lightweight virtualization, and automated lifecycle management.

Microservice decoupling defines certain system functions as "services," and decouples these services from each other. Each service can be deployed separately, making it simple and easy to scale. At the same time, it can be combined flexibly in order to complete new services quickly. The concept of microservice decoupling is very close to the idea of modularizing network functions mentioned at the beginning of this section.

The 5G Core Network architecture draws on the design principle of microservice decoupling to design the service functions of the 5G network (such as mobility management and session management) into independent functional modules, which are coupled weakly with each other and communicate in a service-based manner based on an open API (Application Programming Interface). This weak coupling is based on open APIs (Application Programming Interface), communicated in a service-based manner, and managed with a service governance framework (service module registration, discovery, and orchestration management) to support the rapid launch of new services through the flexible combination and independent upgrade of service-based modules.

The Core Network control plane of the 5G network is a service-based architecture. The 5G network defines several Network Functions (NFs), and the capabilities of each NF are presented to the public by way of services. Specifically, the NFs provide services to any other NFs that are allowed to use them through a service-based interface. The 5G service-based architecture model is shown in figure 3-57.

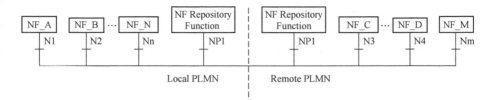

Figure 3-57 The 5G service-based architecture model

The overall 5G network architecture includes the service-based architecture of the Core Network control plane, which is described in section 3.1 of this book.

3.3.2 Key concepts of service-based architecture
The design of the service-based architecture contains the following concepts:

(1) NF (Network Function)
(2) NF service
(3) NF service operation
(4) NF service user
(5) NF service provider

A complete service is represented as Nnf type _NF service _NF service operation. For example, the NF Management service provided by NRF, where one of the service operations is "NFRegister," is fully represented as "Nnrf_ NFManagement_NFRegister."

3.3.3 Main features of the service-based architecture
The main features of the 5G service-based architecture include service management, service discovery, inter-service communication, and service authorization, which are described respectively below.

3.3.3.1 Service management
NF is the minimum deployment granularity for the 5G networks, and one type of NF contains one or more NF instances (NF instances) deployed in the network. When a new NF instance comes online, it needs to be registered with the NF Repository Function (NRF). There are two methods of registration: the NF instance contacts the NRF to register for itself, or the network management

(Operation, Administration, and Management [OAM]) system registers the NF instance in NRF.

The NRF stores information about available NF instances in the network (called the NF profile). When an NF queries information about other NFs, the NRF searches the requested NF type and returns the eligible NF profile. From this point of view, the NRF has a function similar to the DNS query.

The NF profile includes at least the following information:

(1) The NF instance identifier
(2) NF type
(3) PLMN ID: the PLMN network to which the NF instance belongs
(4) S-NSSAI, Network Slice Instance (NSI) identifier, i.e., the slice type supported by the NF instance
(5) FQDN or IP address of the NF corresponding to the NF instance
(6) NF capacity information
(7) NF priority information
(8) NF Specific Service authorization information
(9) The names of NF services supported by the NF instance

The NF instance can update the NF profile to the NRF at any time after successful registration, such as when it supports a new service. Before the NF goes off-line, it registers with NRF and deletes the NF profile.

When the NF instance itself registers with the NRF, the NF management service of the NRF is invoked, as shown in figure 3-58.

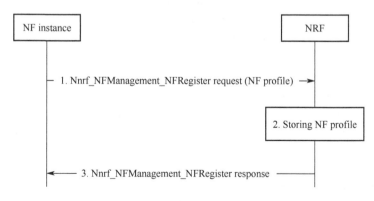

Figure 3-58 The process of the NF instance registering an NF profile with the NRF

Step 1: The NF instance requests the NRF to register its own NF profile.
Step 2: The NRF receives the request and stores the NF profile.

Step 3: The NRF replies to the NF instance with a registration response message, indicating successful registration.

3.3.3.2 Service discovery

Service discovery is when an NF wants to find another NF that provides a certain service. The NF that finds the service is called the NF consumer, and the NF that provides the service is called the NF producer.

There are two implementations of service discovery: either the NF consumer finds a suitable NF producer by means of information about the locally configured NF producer, or the NF consumer gets information about the NF producer by means of a query to the NRF. The flow of the latter scenario is shown in figure 3-59.

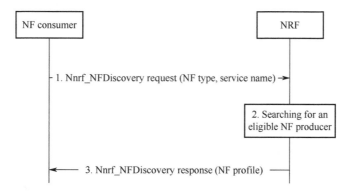

Figure 3-59 NF consumer inquiring NRF and discovers the NF producer

Step 1: The NF consumer requests the NRF to discover the NF producer, providing information such as the NF type and service name.
Step 2: After receiving the request, the NRF finds the eligible NF producer locally.
Step 3: The NRF replies to the NF consumer with a discovery response message, carrying the NF profile of the NF producer.

3.3.3.3 Inter-service communication

In the framework of service-oriented architecture, there are two mechanisms for communication between the NF consumer and NF producer, namely the request-response and the subscription-notification:

(1) Request-response: The NF consumer requests an immediately responsive service from the NF producer. When an NF producer is asked by an NF consumer to provide a specific NF service, the NF consumer may perform an action, may provide information, or may perform both an action and provide information. Sometimes the NF consumer requests a service by

which the NF producer needs to obtain information from other NFs before it can respond, at which point the NF producer becomes the NF consumer and requests services from other NF producers. Request-response interaction between the NF consumer and the NF producer is a request message plus a response message between the two NFs, and the response can be considered an immediate response. Even though the NF producer may have triggered other request-response processes, the response to the NF consumer is still rapid. The request-response communication mechanism is shown in figure 3-60.

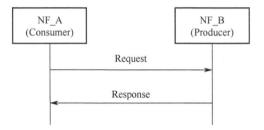

Figure 3-60 The request-response communication mechanism

(2) Subscription-notification: The NF consumer subscribes to a service provided by the NF producer, and the NF consumer sends out a subscription message that includes the subscribed events, subscribed objects, notification trigger conditions, and notification frequency. Notification is the response to the subscription. A notification message is sent by the NF producer, and provides information based on the content of the NF consumer's subscription. A notification can be an instant response to a subscribed message. The subscription-notification communication mechanism is shown in figure 3-61.

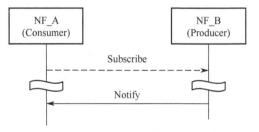

Figure 3-61 The subscription-notification communication mechanism

Another special instance of the subscription-notification communication mechanism is the proxy subscription, where NF A (the proxy consumer) acts as a proxy for NF C (the actual consumer) and subscribes to NF B (the producer). The subscription message will indicate that the notification object is NF C. When the notification condition is met, NF B sends a notification message to NF C. The subscription-notification proxy subscription mechanism is shown in figure 3-62.

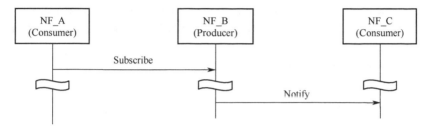

Figure 3-62 The subscription-notification proxy subscription mechanism

3.3.3.4 Service authorization

The purpose of NF service authorization is to check whether the NF consumer is allowed to obtain the services provided by the NF producer. The authorization of services occurs in the process by which the NF consumer discovers the NF producer by inquiring about NRF, and the authorization is performed by the NRF, including the policy of the NF service provider, the policy of the service operator, and the inter-operator agreement.

Service authorization information can be found in the NF profile (of the NF producer), for example, which NF consumers are allowed to discover the NF producer and can further indicate which NF services of the NF producer are allowed to be discovered. The process by which the NF producer authorizes the NF consumer is shown in figure 3-63.

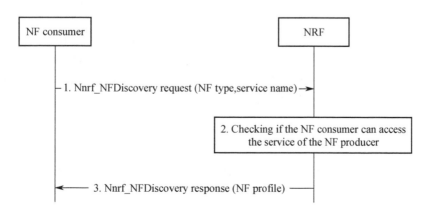

Figure 3-63 NRF queries whether the NF producer authorizes the NF consumer process

Step 1: The NF consumer asks the NRF to discover the NF producer, providing information such as the NF type and service name.

Step 2: After receiving the request, the NRF looks up the eligible NF producer locally, and enquires whether the NF producer authorizes the NF consumer to discover it and its services.

Step 3: Based on the authorization result, the NRF replies to the NF consumer with a discovery response message carrying the NF profile of the NF producer.

3.3.4 Interface protocol for service-based architecture

The protocol stack of the service-based architecture interface is shown in figure 3-64 and is defined in 3GPP TS 29.500 [9].

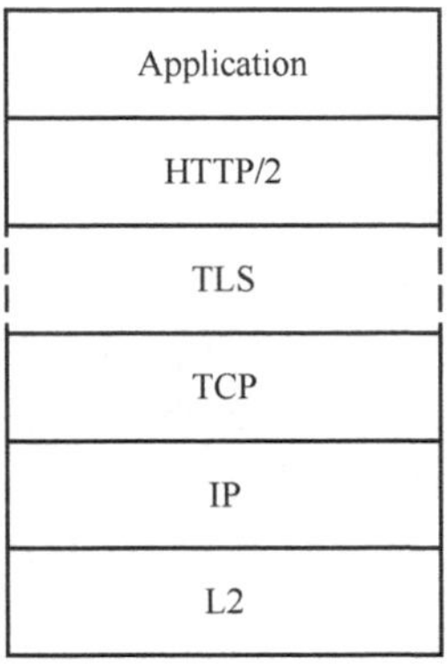

| Application |
| HTTP/2 |
| TLS |
| TCP |
| IP |
| L2 |

Figure 3-64 Protocol stack of the service-based architecture interface

The application layer (Application) is a JSON protocol, compliant with IETF RFC 8259 [10], with extended support for 3GPP-defined parameters.

Below the Application layer is the HTTP/2 protocol, which complies with IETF RFC 7540 [11].

The transport layer protocol is TCP (Transmission Control Protocol) and supports TLS (Transport Layer Security) security encryption. 5G networks require encryption for communication between NFs, and if no other security mechanism is used in the network, the messages must be transmitted with encryption using TLS.

3.3.5 Summary

This section has explained the concept of service-based architecture in 5G networks, which has three major advantages for the 5G Core Networks:

(1) Modularity for easy customization. Each 5G software function is defined by a fine-grained service, which makes it easy to customize and orchestrate the network at the granularity of the service according to service scenarios.

(2) Lightweight and easy to expand. The interface is based on Internet protocols and uses flexible API interactions. Internally, it reduces network configuration and signaling overhead; externally, it provides a unified interface with open capabilities.

(3) Independent and easy to upgrade. Services can be deployed independently and released in grayscale, allowing network functions to be quickly upgraded and new features introduced.

In short, services can be rapidly deployed and flexibly scaled up and down based on the virtualization platform.

3.4 Network slicing

3.4.1 Introduction

Network slicing is one of the key new features of the 5G network architecture, and has a profound impact on the deployment form, architecture design, and service models of the 5G networks.

5G network slicing is driven by both demand and technology.

From the perspective of demand, the target markets for the 5G networks include differentiated services such as enhanced mobile broadband (eMBB), massive Internet of Things (mIoT), and Ultra Reliable and Low Latency Communication (URLLC). These scenarios raise highly differentiated demands on the implementation and deployment of network functions, and it is difficult to accommodate all types of different use cases simultaneously using a traditional single network. If multiple independent physical networks are deployed, the deployment cost for operators will increase significantly. In order to support diverse use cases based on a unified hardware platform, 5G networks need to support logical isolation of multiple dimensions such as service, function, security, transmission, and operation and maintenance, on the basis of shared hardware resources.

From the perspective of technological evolution, Cloud computing, virtualization, and software-based evolution directions are becoming clearer, and have become important foundation technologies for 5G network design. The increasing popularity of network function virtualization (NFV) and software-defined networking (SDN) provides powerful technical support for network function modularization, component orchestration and management, and dynamic configuration and efficient scheduling of network resources.

For these reasons, on a unified network infrastructure, 5G network slicing was born in order to provide mutually isolated network environments for various application scenarios, and to customize network functions and features flexibly, according to their respective needs, guaranteeing the QoS requirements of different services.

3.4.2 Definition and identification of network slices

A network slice is a complete logical network that includes functions such as an Access Network, transmission network, and Core Network.

Network slicing is a technical means to isolate resources and functions such as computing, storage, and transmission in the network. Specific network slices may deploy different functional network elements (e.g., mobility management, billing management, policy control management, and security management), achieve varying performance requirements (e.g., latency, mobility, reliability, and rating), or provide support for specific user groups (e.g., government employees, roaming users, or virtual operators).

A network slice is identified by an S-NSSAI, which includes Slice/Service Type (SST) and Slice Differentiator (SD):

(1) SST: Used to describe the characteristics of a slice in terms of features and services, such as eMBB type and URLLC type.
(2) SD: Optional information, which is a generic identifier to distinguish different network slices with the same SST characteristics. A typical usage scenario is that the SD can be used to indicate the tenant to which the slice belongs.

In roaming scenarios, in order to achieve a consistent understanding of network slice requirements from subscribers across different operators, it is necessary to define some standardized values for network slice types (see table 3-6). Typical 3GPP standardized network slice types are the eMBB slice, URLLC slice, mIoT slice, and V2X slice.

Table 3-6 Standardized slice types

Types of slices/services	SST figure	Attributes
eMBB	1	Supports traditional mobile broadband services and their enhancements
URLLC	2	Supports ultra-low-latency and high reliability services
mIoT	3	Supports massive IoT terminal services
V2X	4	Supports V2X service

If an S-NSSAI contains only standardized SSTs, then the S-NSSAI can identify a standardized network slice. If an S-NSSAI contains both SSTs and SDs or only non-standardized fetched SSTs, then this S-NSSAI can identify a non-standardized network slice. A standardized S-NSSAI can be applied to all PLMNs, and a non-standardized S-NSSAI can only uniquely identify a network slice within a PLMN.

Multiple network slices are identified by Network Slice Selection Assistance Information (NSSAI), which is an assemblance of S-NSSAIs. NSSAI in 5G networks can be Requested NSSAI (the set of network slices that UE requests access to) or Allowed NSSAI (the set of network slices to which the network allows UE access).

In addition to network slices, the concept of network slice instances is also introduced in 5G networks. A network slice instance is a specific instantiation of a deployed network slice. It is important to note that the relationship between network slices and network slice instances is not a one-to-one correspondence. Depending on the operator's operational or deployment needs, network slices and network slice instances are in a many-to-many mapping relationship; that is, a network slice instance can be associated with one or more S-NSSAIs, and an S-NSSAI can be associated with one or more network slice instances. Multiple network slice instances associated with the same S-NSSAI can be deployed in the same tracking area or in different tracking areas.

In principle, it is preferable that the network elements between different slices are deployed independently to achieve the best slice isolation. However, a UE can have access to one or more network slices at the same time due to the complexity of the service requirements of certain UEs. Because most of the mobility management attributes of a UE, such as location and accessibility, are naturally controlled at the granularity of a UE rather than a slice, this part of the network functionality then needs to be shared across slices. This deployment requirement also directly influences the design of 5G system architecture.

In the scenario of network slice deployment, there are certain differences between 5G network element functions and slice deployments:

(1) Proprietary functions of network slices. For example, the session management network element SMF and the user-plane function UPF are responsible for establishing PDU sessions on a specific network slice instance, so SMF and UPF can be deployed as network slice proprietary functions, thus achieving mutual isolation of SMF and UPF between different network slices.

(2) Network slice-sharing function. As mentioned above, for smart UEs that support multi-slice access capability, in order to realize the complexity of unified access control and mobility management for such UEs, these multiple network slices need to share the same AMF network element to provide services for that UE. Similarly, UDMs responsible for UE contract

management and PCFs controlling UE granularity policies may be shared among slices.

(3)   PLMN granularity network element. The 5G network introduces the Network Slice Selection Function (NSSF) as a centralized deployment element at PLMN granularity on the core side to achieve a flexible selection of network slices for UE access. The NSSF senses the slice capacity of each AMF in the PLMN, and also dynamically adjusts the availability of network slices at TA granularity according to operator requirements, network congestion, or local policies on demand.

(4)   Multi-granularity deployment of network elements. The functions of some network elements have multiple granularities, and operators can deploy them according to different granularities (e.g., PLMN granularity, slice sharing, and slice proprietary) to maximize the isolation of slices. Using NRF as an example, 5G networks support three granularities of NRF network element deployment:

a.   PLMN granularity NRF network elements: responsible for discovering PLMN granularity network elements or cross-PLMN network elements.

b.   NRF network elements at slice-sharing granularity: responsible for the discovery of network elements at slice-sharing functionality, such as the discovery and selection of AMF network elements.

c.   NRF network element at slice-exclusive granularity: responsible for discovering the network element within a network slice instance. If an AMF network element needs to discover a network element within a network slice instance corresponding to an S-NSSAI, such as a PDU session establishment process, the AMF network element can request a slice-granularity NRF network element to discover the target network element.

3.4.3 UE support for network slicing

To realize end-to-end slicing implementation, the UE can store the slicing configuration information (Configured NSSAI) or Default Configured NSSAI applicable to a particular PLMN to assist in the network slicing selection process during the initial network entry of the terminal. For a VPLMN, the UE can also store the Configured NSSAI applicable to that PLMN and the HPLMN S-NSSAI with which each of the S-NSSAIs has a mapping relationship.

The Configured NSSAI identifies the set of S-NSSAIs that the UE can request at a particular PLMN. In general, the Configured NSSAI of the HPLMN stored on the UE is consistent with the signed S-NSSAI of the UE stored on the network side.

In addition to the Configured NSSAI, the network issues the corresponding UE routing strategy to the UE, based on the signed information of the UE in the registration or UE configuration update process. The URSP includes the Network Slice Selection Policy (NSSP), which is used by the UE to sense the association of an application with a network slice and assist the UE in the subsequent admission and session establishment process to achieve end-to-end network slice selection. In addition, the S-NSSAI value contained in the NSSP is the S-NSSAI of the HPLMN.

When the contracted S-NSSAI of the UE is modified, the Core Network updates the above-mentioned UE Configured NSSAI and URSP as needed.

The UE indicates the slice that the network itself wants to access through the Requested NSSAI. The network side selects the appropriate AMF network element, network slice set, and network slice instance (NSI) for the UE based on the Requested NSSAI, and finally returns the Allowed NSSAI to the UE in the registration reply message. The 3GPP TS 23.501 [1] standard specifies that a UE can request access to up to eight slices at the same time, thus the Requested NSSAI and Allowed NSSAI contain up to eight S-NSSAIs.

3.4.4 Key processes for network slicing

The UE accesses and uses a particular slice for data transmission, comprising two processes:

(1) The UE achieves registered access to one or more network slices through a registration process, i.e., registration access to a network slice.
(2) The UE establishes a PDU session in a specific network slice, i.e., session establishment within a network slice.

3.4.4.1 Registration access for network slices

During the process of initial registration or mobility registration, the UE needs to complete the registration of one or more network slices. This process consists of determining the slice or slices (Allowed NSSAI) that the UE is allowed to access in that registration area and select the appropriate AMF that can serve that network slice or slices.

In Rel-15/16, the slice deployment needs to ensure that the slice support capability is the same in the UE's registration area. That is, in general, the UE does not need to consider whether the Allowed NSSAI will change until it is moved out of the current registration area. It should be noted that the determination of the Allowed NSSAI is also based on the access technology, since the registration process and registration area of 3GPP access technology and non-3GPP access technology are handled separately. Figure 3-65 shows the registration access process for network slicing.

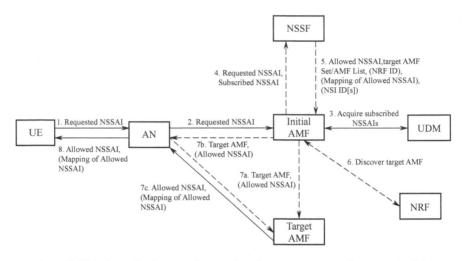

Figure 3-65 Schematic diagram of the registration access process for network slicing

Step 1: Determination of Requested NSSAI. When a UE registers to a PLMN, the UE can carry the Requested NSSAI at both the Access Stratum (AS) and NAS layers to indicate the set of network slices to which the UE requests access. The UE can flexibly determine the Requested NSSAI according to various scenarios. For example, the Requested NSSAI can be one of the following options:

(1) When the UE stores the Allowed NSSAI corresponding to the access type used to send the Requested NSSAI, the UE uses this Allowed NSSAI or a subset of it as the Requested NSSAI for this time.

(2) When the UE stores the Allowed NSSAI corresponding to the access type used to send the Requested NSSAI, the UE uses the Allowed NSSAI or a subset thereof, plus one or more S-NSSAIs, as the current Requested NSSAI. Among these, one or more S-NSSAIs are those included in the Configured NSSAI but are not included in the Allowed NSSAI as S-NSSAI.

(3) When the UE does not have a corresponding Allowed NSSAI for this PLMN, the UE uses the Configured NSSAI corresponding to this PLMN or a subset thereof as the Requested NSSAI for this time.

(4) When the UE does not have the Configured NSSAI corresponding to this PLMN nor the Allowed NSSAI, the UE takes the Default Configured NSSAI as the Requested NSSAI for this time.

Step 2: The Access Network selects the correct AMF network element based on the Requested NSSAI provided by the AS and the temporary identification of the UE. During the selection process, temporary identifiers contained in the AS are given higher priority than Requested NSSAI. If the UE carries both the Requested NSSAI and the temporary identifier in the AS, the Access Network element gives priority to the AMF based on the temporary identifier. If the appropriate AMF cannot be selected based on the temporary identifier (e.g., if the Access Network element cannot connect to the AMF indicated by the temporary identifier), then the AMF selection is made based on the Requested NSSAI. If the UE carries neither the Requested NSSAI nor the temporary identity when establishing a connection to the Access Network, the Access Network element forward the UE's registration request to the default AMF network element.

Step 3: The initial AMF obtains the user signing S-NSSAI information from the UDM.

Step 4: After the initial AMF network element selected by the Access Network receives the Requested NSSAI requested by the UE, if this AMF can satisfy the Requested NSSAI, then the AMF further forward the UE's request based on the Requested NSSAI requested by the UE, the current location of the UE, the signed slice type, the PLMN to which the UE belongs, and other access types corresponding to the Allowed NSSAI. If the AMF cannot satisfy the Requested NSSAI, then the AMF network element queries the NSSF network element to request the NSSF network element to determine the Allowed NSSAI and AMF for the UE. The NSSF network element address can be configured on the AMF network element.

Step 5: The NSSF determines the Allowed NSSAI corresponding to the current access technology, the Mapping of Allowed NSSAI, and information like the AMF Set, which can serve the Allowed NSSAI or target AMF network element, based on comprehensive consideration of elements such as the Requested NSSAI sent by the AMF network element, the current location of the UE, the type of slicing contracted, the PLMN to which the UE belongs, the Allowed NSSAI corresponding to other access types, and the slicing capability of the AMF network element. Based on the operator configuration, the NSSF can also determine the NRF network element corresponding to each S-NSSAI at the network slice granularity for network element discovery within the network slice during the registration process. The NSSF sends the network slice selection result to the current AMF network element.

Step 6: If the NSSF sends the target AMF network element in which the target AMF network element is located to the current AMF network element, this

AMF network element asks the NRF network element to discover the target AMF network element based on the target AMF Set.

Step 7: The initial AMF network element supports triggering the redirection of the UE to the target AMF based on the result of the network slice selection. Considering the varying isolation between slices, the AMF redirection process can be implemented through direct interaction between two AMFs or through the Access Network.

Step 8: The AMF determines the registration area of the UE based on the Allowed NSSAI to ensure that any S-NSSAI in the Allowed NSSAI can be accessed within the registration area. The AMF sends information such as the Allowed NSSAI, the Mapping of the Allowed NSSAI, the URSP information obtained from the PCF network element, the registration area, and the Rejected S-NSSAIs together to the UE. The AMF also notifies the Access Network of the Allowed NSSAI through an N2 message. The Access Network side senses the Allowed NSSAI in order to achieve differential scheduling and management of sliced resources.

3.4.4.2 Mapping of slices to slice instances

In the network slice selection process, the many-to-many mapping relationship between network slices and network slice instances can be supported. In order to solve the problem of multiple network slice instance deployment, two ways (Early Binding and Late Binding) of network slice instance selection are specified in the standard process. Accordingly, the AMF network element can obtain the network slice instance identification (NSI ID) corresponding to the S-NSSAI from the NSSF network element in the registration process or PDU session establishment process.

(1) Early Binding: In the registration process, the NSSF selects the NSI ID corresponding to each S-NSSAI in the Allowed NSSAI for the UE, as well as determines the corresponding target NRF network element (slice proprietary granularity) for the selected network slice instance to support the selection of network elements within the network slice instance in the subsequent PDU session establishment.

(2) Late Binding: In the registration process, the NSSF only determines the Allowed NSSAI for the UE. When the UE first initiates the PDU session establishment process associated with an S-NSSAI, the NSSF determines the NSI ID corresponding to the S-NSSAI and the corresponding target NRF network element (slice proprietary granularity) to complete the network element selection within the network slice instance during PDU session establishment.

3.4.4.3 Session establishment within a network slice

After the registration process, the UE can initiate a PDU session establishment process to carry the data flow of the service through the network slice instance. Each PDU session is associated with an S-NSSAI and a DNN, where the S-NSSAI corresponding to the PDU session must be one of the S-NSSAIs in the UE's Allowed NSSAI. The UE will determine the S-NSSAI corresponding to the PDU session as well as the DNN based on the URSP policy or local configuration, and the S-NSSAI contained in the Allowed NSSAI in combination. If the UE cannot determine the S-NSSAI corresponding to the application, then in the PDU session establishment process, without the UE specifying the S-NSSAI, the network side will determine the slice corresponding to the session at this time.

According to the service requirements of the UE, a UE can establish multiple PDU sessions associated with different network slices at the same time, where the session user-planes in different network slices are isolated from each other. The same PDU session cannot be migrated between different network slices within a PLMN.

The basic principle of network element discovery within a network slice instance is based on the S-NSSAI of each PLMN, while the NRF within that PLMN discovers the network elements within the network slice instance.

For the UE request to establish the S-NSSAI carried by the PDU session, if the AMF has already obtained the NSI ID corresponding to each S-NSSAI in the Allowed NSSAI and the target NRF network element corresponding to each NSI ID in the registration process, then the AMF can directly request the NRF network element based on the NSI ID corresponding to the S-NSSAI carried by the UE, and the target address of the NRF network element can request the discovery of the SMF network element.

If the AMF cannot determine the NRF network element used for network element discovery within the network slice instance, then the AMF first requests the NSSF network element to determine the NSI ID corresponding to this S-NSSAI, and the target NRF network element corresponding to the NSI ID, to determine the SMF network element corresponding to this PDU session establishment through the NRF.

3.4.4.4 Roaming scenario support

In the roaming scenario, the end-to-end network slicing is implemented by the home domain slicing and the visited domain slicing together. In order to simplify the interaction between operators, the network slice mapping between the home domain and the visited domain is determined based on the roaming agreement.

When the UE initially registers to the network slice, the UE determines the mapping relationship of the S-NSSAI in the Requested NSSAI to the S-NSSAI of the HPLMN based on the configuration information and provides this mapping relationship to the network. In the network slice selection process, not only the Allowed NSSAI of the UE is determined but also the mapping relationship between the S-NSSAI in the Allowed NSSAI and the S-NSSAI of the HPLMN is determined based on the roaming protocol. Based on the operator's policy and configuration, the network slice mapping relationship between the attribution domain and the visiting domain can be determined by AMF or NSSF, while the Allowed NSSAI is determined by the visiting domain NSSF or AMF.

During the PDU session establishment process, after the UE determines the HPLMN S-NSSAI corresponding to the application according to the rules of URSP, the UE also needs to determine the VPLMN S-NSSAI corresponding to the HPLMN S-NSSAI according to the locally configured network slice mapping relationship between the attribution domain and the visiting domain. The VPLMN S-NSSAI will also contain in the Allowed NSSAI acquired by the UE from the network.

In the roaming scenario, the discovery of the network elements in the attributed and visited domains is achieved through the NRF deployed by this PLMN. The SMF selection in the Home Routed scenario includes two parts, namely the VPLMN SMF element selection and HPLMN SMF element selection, as shown in figure 3-66, where the VPLMN SMF element selection can refer to the VPLMN SMF in the LBO scenario. The HPLMN SMF selection mechanism depends on the roaming agreement signed between operators and the deployment of NSSF network elements and is implemented in two ways:

(1) If the VPLMN and HPLMN deploy NSSF network elements at the same time, the HPLMN SMF selection is jointly implemented by the NSSF network elements and NRF network elements of different PLMNs.

   If the AMF is unable to determine the NRF network element used to perform network element discovery within the HPLMN network slice instance, then the AMF first requests the HPLMN NSSF network element through the VPLMN NSSF network element to determine the HPLMN NSI ID corresponding to this HPLMN S-NSSAI and the target HPLMN NRF network element corresponding to this HPLMN NSI ID. Here, the HPLMN S-NSSAI is the HPLMN S-NSSAI associated with this PDU session. Based on the HPLMN S-NSSAI and the corresponding HPLMN NSI ID, the AMF requests the HPLMN NRF network element via the VPLMN NRF network element to determine the HPLMN SMF network element in the HPLMN network slice instance.

(2) If the HPLMN does not deploy the NSSF network element, or if there is no interface between the NSSFs of the two PLMNs, the AMF network element

of the VPLMN can obtain the HPLMN NRF network element address to discover the HPLMN SMF network element through local configuration.

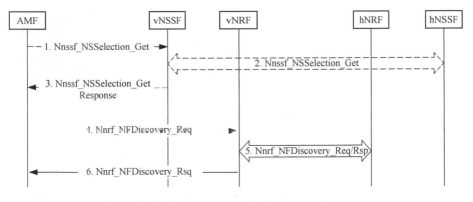

Figure 3-66 SMF selection in the home routed scenario

3.4.5 Summary

As a key feature of 5G networks, network slicing provides a key technical foundation for operators to build isolable and customizable logical networks. The network slicing mechanism in 5G networks includes the definition and identification of network slices, as well as the features required by UEs and network sides to support network slicing access and control. The main processes involved in network slicing include the slice request and selection process in the registration process, the slice-related network element selection process in the session establishment process, the mapping relationship between slices and slice instances, and how to realize the mapping of two sets of slices in the visiting and belonging domains in the roaming scenario.

3.5 Edge computing

3.5.1 Introduction

In 4G and previous traditional mobile network architectures and deployments, user-plane devices were deployed in a tree topology. uplink subscriber traffic passes through the base station and backhaul network, and finally accesses the data network through centrally deployed anchor gateways, as shown in figure 3-67. These anchor gateways are generally deployed at higher locations in the network, such as regional central server rooms. In this deployment method, service message traffic is concentrated on a few gateways and external outlets, and the network topology is relatively simple, which is convenient for operators to conduct centralized service control and message processing at the anchor points.

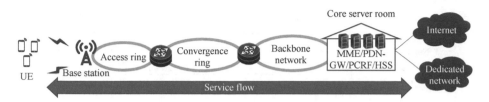

Figure 3-67 Centralized 4G deployment

With the widespread deployment of 4G networks, the mobile Internet has been a huge success, becoming one of the main ways for users to get online, and leading to an explosive growth in mobile service traffic. The traditional centralized anchor deployment approach is proving increasingly incapable of supporting this fast-growing mobile service traffic model. In a network with centralized anchor gateway deployment, growing traffic is eventually concentrated at the gateway and core server room, placing ever higher demands on backhaul network bandwidth, server room throughput, and gateway specifications. Meanwhile, the long backhaul network distance from the Access Network to the anchor gateway and the complex transmission environment also leads to a greater delay and jitter in user message transmission.

In these contexts, the industry has proposed the concept of Edge Computing (EC). By moving user-plane network elements and service processing capabilities down to the edge of the network, edge computing enables distributed local processing of service traffic, avoiding excessive centralization of traffic and thus significantly reducing the specification requirements for core server rooms and centralized gateways. Edge computing also shortens the distance of the backhaul network and reduces the end-to-end transmission delay and jitter of user messages, making the deployment of ultra-low-latency services possible.

There is no clear industry consensus on the exact location of the network edge to which edge computing refers. A schematic diagram of edge computing deployment is shown in figure 3-68. It is important to note that as the location of downward deployment decreases, the deployment conditions, room environment, and resource utilization efficiency will become worse. Therefore, edge computing is not a case of "the lower, the better," but should strike a certain balance between user experience and deployment cost. For example, gateways and services are deployed down from large regional centers to the edge of metro networks to cover one or more urban areas.

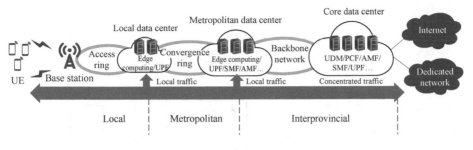

Figure 3-68 Edge computing deployment

3.5.2 International standards for edge computing

Edge computing has been a hot topic in the industry since the discussion of control-and-forward separation for 4G. In terms of early standardization, ETSI established the Mobile Edge Computing (MEC) working group in December 2014 to carry out standardization for edge computing platforms [12]. The ETSI MEC system reference architecture is shown in figure 3-69.

In addition to the standardization work of the ETSI MEC ISG, considering the possibility of widespread deployment of edge computing in 5G low-latency and large bandwidth scenarios, 3GPP has also included support for edge computing deployment scenarios since the beginning of the 5G network architecture design. 5G network architecture support for edge computing is realized through the following aspects:

(1) Anchor point switching based on SSC mode 2 and SSC mode 3 (see section 3.2.2.3 of this book). Through the synchronous migration of anchor points and UEs, user services can access the local servers deployed on the edge computing platform through the nearest anchor gateway, thus enabling the transmission support of the session to the edge computing platform.

(2) Service triage scheme based on an Uplink Classifier (UL CL) and Branching Point (BP). In addition to the session anchor point UPF switching defined by the SSC mode, the 5G system also supports the ability to access the edge computing platform for a specific service flow in the same session through a splitting technique, as described in section 3.5.3 of this book.

(3) Interaction between EC and network. The EC acts as an AF and collaborates with the SMF and NEF to achieve synchronous migration of UPF and applications through an event escalation mechanism.

(4) Local Area Data Network (LADN). A special class of data network that provides only a limited range of coverage.

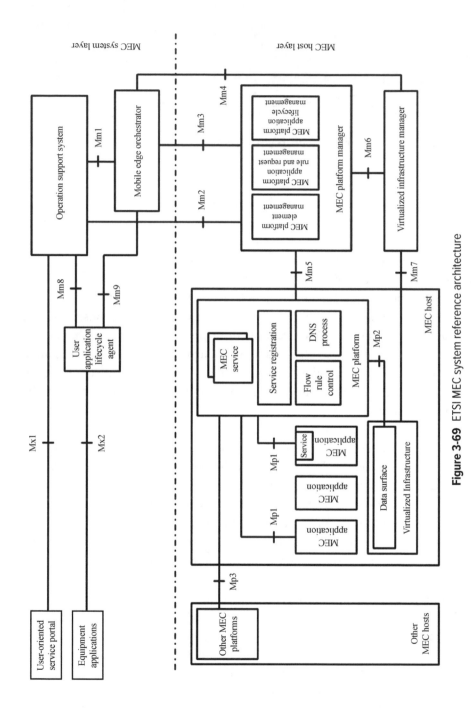

Figure 3-69 ETSI MEC system reference architecture

3.5.3 5G network support for edge computing

3.5.3.1 Local service triage

The main purpose of local service triage is to triage certain services that can terminate in the local network from a lower location to a locally deployed DN based on the characteristics of the services while maintaining a centralized anchor point, thus being able to reduce the end-to-end latency of these services and reduce the load on the backbone network.

Local service triage can be achieved in two ways:

(1) Service triage based on UL CL. UL CL is a type of UPF network element deployed between the AN and the anchor UPF that classifies the uplink messages sent by the UE according to the uplink message characteristics (such as destination IP and port number) provided by the SMF, and forward the uplink messages that match the characteristics directly to the locally deployed DN. Since the UE side always uses the IP address assigned by the anchor UPF, the UL CL-based traffic splitting is transparent to the UE. A local anchor may also be deployed between the UL CL and the locally deployed DN to implement the interface with the locally deployed DN.

(2) BP-based service splitting (which is also a UPF network element deployed between the AN and the anchor UPF) is implemented based on the source IP address of the messages sent by the UE. Therefore, in the BP-based service triage scheme, the locally triaged service flow will use a specific source IP address in the uplink. The BP scheme is only used for IPv6 sessions.

The two traffic splitting schemes based on UL CL and BP are shown in figures 3-70 and 3-71. It should be emphasized that the locally deployed DN and the remote DN in the figures are actually the same DN deployed in different locations. Their DNNs are the same.

Usually, the insertion, deletion, and movement of UL CL/BP are caused by UE movement or service triggering. Although UL CL/BP is also a type of I-UPF, it should be noted that the insertion, deletion, and movement of UL CL/BP are different from the corresponding operation flow of basic I-UPF. The above operations of I-UPF will be completed directly in the processes of UE switching and service request, but these operations of UL CL/BP are often accompanied by the insertion, deletion, and movement of local anchor points. Therefore, SMF will execute the corresponding UL CL/BP operations independently after the corresponding trigger processes (e.g., UE switching and service request) are completed.

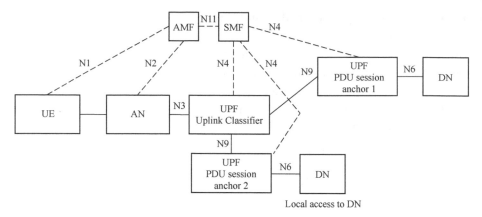

Figure 3-70 Local diversion scheme based on UL CL

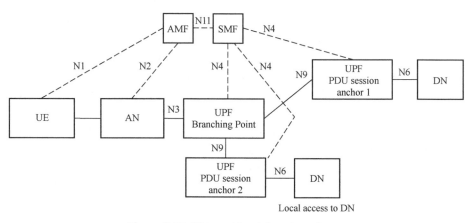

Figure 3-71 BP-based local diversion scheme

3.5.3.2 Interaction of edge computing systems with 5G networks

Due to reasons such as latency and deployment, there is often some synergy between edge computing systems and 5G networks to support features such as end-to-end routing and mobility optimization. In existing protocols, the interaction between edge computing systems and 5G networks is achieved through AF.

However, it is important to note that in an EC deployment environment, there is no standard conclusion as to which part of the EC system the AF specifically maps to.

AF offers the following two main functions:

(1) Providing information related to the local routing of service flows to the 5G network via NEF or PCF to support the SMF in selecting a more appropriate

UPF, performing route forwarding and anchor switching. This information includes the following:

a. identification information of the application service flow, such as DNN, S-NSSAI or application identification, and service IP 5-tuple
b. information on the service routing requirements of the N6 interface, such as routing policy identification or routing tunnel
c. the location where the application routing information applies
d. UE identification, which may be for a single UE or for a group of UEs
e. the possibility of application support migration
f. the effective time and effective area
g. subscription information for session events
h. a UE IP address holds the indication, to indicate whether the application expects the UE IP address to remain unchanged
i. an indication of whether to wait for the AF response

(2) Subscribing to Early Notification and Late Notification. These are the two notification methods for the SMF to send notification events to the AF. The main purpose of this notification event is to notify the UPF Relocation event to the edge computing system, so that the edge computing system can trigger the migration of the application synchronously. Early Notification occurs before UPF migration, while Late Notification occurs after UPF migration is completed or before the UPF path takes effect.

The notification mechanism between the AF and the network is shown in figure 3-72. After the SMF selects the target UPF for migration, if the AF subscribes to Early Notification, the SMF will immediately send a notification event to the AF

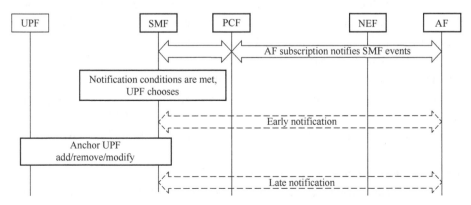

Figure 3-72 Notification mechanism between the AF and the network

and decide whether to execute the subsequent UPF migration immediately or wait for the AF to answer based on whether the AF's subscription event contains an indication to wait for the AF to answer then perform the migration. If the AF subscribes to Late Notification, the SMF decides whether to notify the AF and wait for an answer before the new UPF path is activated or to notify the AF after the UPF migration is completed, based on the Waiting for AF Answer indication.

3.5.3.3 Local Area Data Network (LADN) support
The LADN is a special class of DN that is only accessible within a specific location area, e.g., malls, stadiums, and factories. However, it is worth noting that while edge computing can be used to deploy LADNs, LADNs themselves are not strictly tied to edge computing. For example, LADNs can be implemented by networks that are deployed in centralized locations, while edge computing can also provide wide-area service coverage, not just DNs that are accessible in a local area.

Compared with a normal DN with wide-area coverage, a LADN requires the UE and the network to provide the following capabilities: (1) UE awareness of LADN information to avoid initiating PDU session requests out of the LADN coverage, and (2) network control of PDU sessions based on LADN information and the UE location.

3.5.3.3.1 UE awareness of LADN information
For a specific LADN deployment, the LADN DNN and LADN coverage are configured on the AMF located within the LADN coverage area.

When the UE accesses the 5G network through the registration process, the UE can also carry LADN indication information in the registration request to indicate which LADN DNNs the UE wishes to obtain corresponding LADN information. The AMF determines the LADN information to be sent to the UE based on the UE's location, configuration, and signing information. The LADN information includes the LADN DNN and the LADN coverage area. In addition, the LADN coverage area sent to the UE is the intersection of the LADN coverage area configured by the AMF and the UE registration area.

When the deployed LADN network is updated, the AMF can update the LADN information on the UE through the UE Configuration Update (UCU) process.

The UE determines whether it is currently within the LADN coverage area based on the acquired LADN information. If it is not in the LADN coverage area, the UE will not initiate the establishment of the corresponding PDU session.

3.5.3.3.2 Session management based on LADN information
The network can sense whether the UE is within the LADN coverage area in two ways.

1. EVENT SUBSCRIPTION AND EVENT NOTIFICATION

In the session establishment process, if the SMF determines that the DNN accessed by the UE is a LADN DNN, then the SMF can send an event subscription request to the AMF to subscribe to the UE moving into or out of the LADN coverage area event. When the AMF detects that the UE moves into or out of the LADN coverage, the AMF sends an event notification to the SMF, which carries the event result information.

There are three types of event result information: The UE moving into LADN coverage, the UE moving out of LADN coverage, and the UE location remaining unknown. The latter refers to the fact that the AMF may not be able to determine whether the UE is within the LADN coverage area when the UE is in the idle state.

2. NETWORK CONTROL OF LADN PDU SESSIONS

The SMF can control the LADN PDU session accordingly after it is informed that the UE has moved into the LADN coverage area, or if the UE has moved out of the LADN coverage area, or if the UE location is unknown:

(1) When the UE moves into the LADN coverage area, if the user-plane connection of the LADN PDU session is in the deactivated state, the SMF needs to instruct the UPF to send a data notification message to the SMF when the downlink data arrives, in order to trigger the service request process.
(2) When the UE moves out of the LADN coverage area, the SMF can release the LADN PDU session, or selectively deactivate the user-plane connection and turn off the upload of the UPF data notification message.
(3) When the location of the UE is unknown, the behavior of the SMF is not standardized, but in order to avoid packet loss, the same treatment can be applied to the LADN PDU session as when the UE moves into the LADN coverage.

3.5.4 Summary

Edge computing is one of the key technologies through which 5G networks can support high-traffic and low-latency services. 5G networks provide edge computing capabilities, including local service triage, service and session continuity support, LADN support, and AF and network collaboration enhancement. Through these capabilities, the 5G network can realize part of the service directly carried on the local edge computing platform, thus reducing the service delay of users and the traffic load of the backbone network.

3.6 Network intelligence

3.6.1 Introduction
In recent years, with the growing maturity of machine learning algorithms and breakthroughs in computer hardware computing power, the outcomes of big data and artificial intelligence technologies have been significantly enhanced, providing unlimited possibilities for achieving 5G network intelligence.

During the Rel-15 phase in February 2017, the 5G network architecture introduced the Network Data Analytics Function (NWDAF) and defined the results of network element load data analysis but did not involve the 5G smart network framework.

Therefore, in the Rel-16 phase in April 2017, 3GPP SA2 set up specifically the enabler of Network Automation for 5G (eNA), systematically providing a smart 5G network architecture and richer application scenarios.

The 5G Smart Network Architecture Rel-16 3GPP standard protocol TS 23.288 [13] was officially released in June 2019, comprising an overall framework based on NWDAF, key processes, and the data analysis results that NWDAF can provide.

3GPP focuses on the standardization of the architecture and interface aspects of smart frameworks. Artificial Intelligence (AI)-specific algorithms and training are not included in the 3GPP standardization and discussion. However, to facilitate readers' understanding, this section also gives examples of how NWDAF obtains the corresponding data analysis based on the input data and on what algorithms.

3.6.2 Overall structure of smart network architecture
This section explains the generic NWDAF-based smart network architecture.

The 5G smart network architecture based on the NWDAF is shown in figure 3-73. From the logical architecture, as part of the 5G network architecture, the NWDAF can realize data collection and data analysis feedback by interacting with the Core Network elements and OAM network elements:

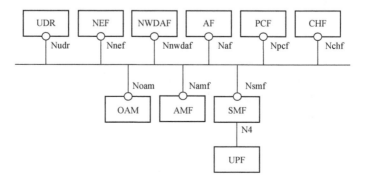

Figure 3-73 5G smart network architecture based on the NWDAF

(1) Data collection: based on event subscription, the NWDAF can collect data from the Core Network element and OAM.

(2) Feedback on data analysis: the NWDAF can distribute the results of data analysis to different network elements (e.g., PCF and OAM) according to demand.

From a deployment perspective, a single NWDAF instance or multiple NWDAF instances can exist in the same PLMN. 5G network architecture can support NWDAF deployment as a central functional network element, or NWDAF deployment as a distributed functional network element, or a combination of both. An NWDAF instance can be implemented as a sub-function of a particular network element.

The data collection schematic is shown in figure 3-74. Based on event subscription, the NWDAF can collect data from the 5G Core Network elements through the Nnf service-based interface. In addition, the NWDAF can collect data from the OAM by invoking the OAM service.

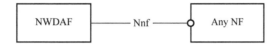

Figure 3-74 Data collection

When the NWDAF collects data from different network elements, the services in use are shown in table 3-7.

Table 3-7 Services used by the NWDAF when collecting data from different network elements

NE	Service
AMF	Namf_EventExposure
SMF	Nsmf_EventExposure
PCF	Npcf_EventExposure
UDM	Nudm_EventExposure
NEF	Nnef_EventExposure
AF	Naf_EventExposure
NRF	Nnrf_NFDiscovery
	Nnrf_NFManagement

Also, how the NWDAF collects data from the UPF has not yet been defined in Rel-16. It will be discussed in Rel-17.

3.6.3 Key processes for network intelligence

This section explains the general data collection process and the feedback process of analysis results.

NWDAF can collect data from the control plane, network management, third-party AF, and RAN. The common data analysis result feedback process will be described according to two dimensions: 5G network elements and third-party AFs.

For RAN data collection, the Rel-16 standard only supports collection through the network manager. Since the data of the same user is distributed on different network elements, for data analysis, the following section describes how to correlate the data on different network elements.

3.6.3.1 The data collection process

3.6.3.1.1 Collecting data from the NFs/AF

The process by which the NWDAF collects data from NFs/AF is shown in figure 3-75, based on event subscription.

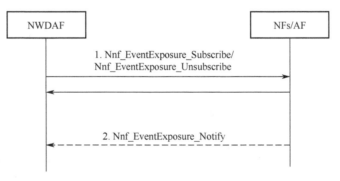

Figure 3-75 The process by which NWDAF collects data from NFs/AF

Step 1: Through the Nnf_EventExposure_Subscribe or Nnf_EventExposure_Un-subscribe service-based functional network element, the NWDAF subscribes or unsubscribes one or several events (via Event ID), so that it gathers data from the appropriate functional network element.

Step 2: If the functional network element has prepared the appropriate data for the event subscribed by the NWDAF, this functional network element can provide data about this event for the NWDAF via Nnf_EventExposure_Notify.

3.6.3.1.2 Collecting data from a third-party AF

The process by which the NWDAF collects data from a third-party AF is shown in figure 3-76, based on event subscription.

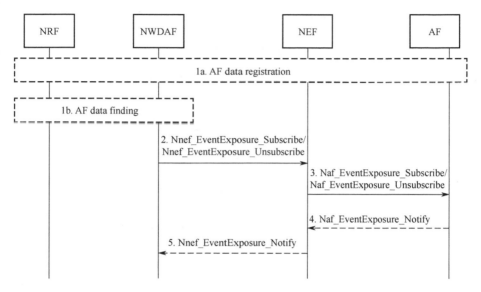

Figure 3-76 The process by which the NWDAF collects data from a third-party AF

Step 1a: AF will equip the NEF with data supported by each service; the NEF can produce a NEF Profile (Event ID, AF ID, and Application ID) and register to the NRF.

Step 1b: When the NWDAF needs to collect data from a third-party AF, it first queries the NRF to obtain the appropriate NEF address.

Step 2 to step 5: The NWDAF subscribes or unsubscribes data from a third-party AF in the NEF. Based on a request from the NWDAF, the NEF further subscribes or unsubscribes AF data.

3.6.3.1.3 Collecting data from the OAM

The NWDAF can gain data from the following (but not limited to) types of network management data from OAM:

(1) NG-RAN or 5GC Performance Measurement (PM) data
(2) 5G end-to-end Key Performance Indicator data
(3) General Performance Assurance (PA) data
(4) Performance Management data
(5) Fault Supervision (FS) data

The process by which the NWDAF collects data from the OAM is shown in figure 3-77.

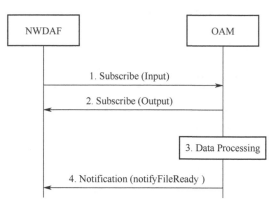

Figure 3-77 The process by which the NWDAF collects data from the OAM

Step 1 and step 2: The NWDAF subscribes network management data from the OAM; the OAM notifies NWDAF's success of subscription.

Step 3 and step 4: The OAM prepares the network management data, and the OAM notifies the NWDAF that the network management data is ready and the data storage address. Based on the vendor's implementation, the NWDAF can fetch OAM data from the storage address via FTP.

3.6.3.1.4 Data connection

Since the data of the same user is distributed on different network elements (e.g., location information on the AMF, session information on the SMF, policy information on the PCF, radio channel information on the RAN, flow information on the UPF, and service information on the AF), the NWDAF needs to analyze all of the user's end-to-end data (e.g., user service experience. See section 3.2 of this book). The NWDAF correlates the data on different network elements with different association marks, as shown in figure 3-78.

After the NWDAF has collected data from each network element, it needs to connect it together in order to undergo data model training. The data collected by the AWDAF can be at the UE level, session level, QoS Flow level, or stream of service level. Therefore, it is necessary to clarify how data of different granularity uploaded by each network element is connected, and which kind of link identifier is used. For the link identifiers on each network element, see table 3-8.

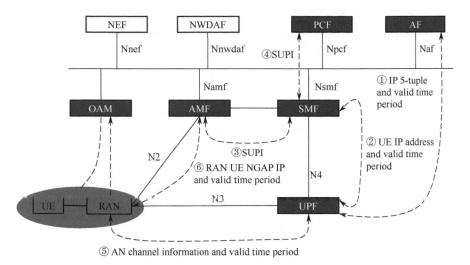

Figure 3-78 NWDAF's two-by-two correlation of data on different network elements

Table 3-8 Link identifiers on each network element

Link Identifiers	Usage
IP 5-tuple and valid time period	Used to link AF data and UPF data
AN channel information and valid time period	Used to link UPF data and OAM data from RAN (e.g., RSRP, RSRQ, SINR data, as is described in section 3.6.3.2 of this book)
UE IP address and valid time period	Used to link UPF data and SMF data
SUPI	Used to link SMF data and AMF data
SUPI, DNN and S-NSSAI or UE IP Address	Used to link SMF data and PCF data
RAN UE NGAP ID and valid time period	Used to link AMF data and OAM data from RAN (e.g., RSRP, RSRQ, SINR data, as described in section 3.6.3.2 of this book)

3.6.3.2 Feedback process for data analysis results

The feedback process for data analysis results is shown in figure 3-79. The NWDAF can provide data analysis results to the 5G Core Network elements as well as the OAM through the NWDAF service-based interface. The specific data analysis results are described in section 3.6.4 of this book.

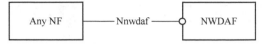

Figure 3-79 Feedback of data analysis results

The services provided by the NWDAF in offering data analysis results to different 5G network elements are shown in table 3-9.

Table 3-9 Services used for NWDAF data collection

NE	Service
NWDAF	Nnwdaf_AnalyticsSubscription
	Nnwdaf_AnalyticsInfo

3.6.3.2.1 Subscription/notification mode
3.6.3.2.1.1 Feedback of data analysis results to network elements

The subscription and notification of data analysis results initiated by the network element is shown in figure 3-80. Based on the subscription/notification model, the NWDAF service consumers (e.g., Core Network elements and OAMs) subscribe to the flow of data analysis results to the NWDAF through the Nnwdaf_AnalyticsSubscription service operation.

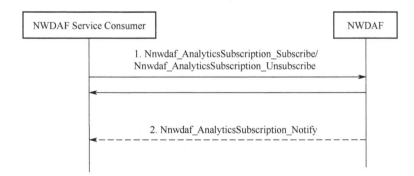

Figure 3-80 Subscription and notification of data analysis results initiated by the network element

Step 1: The NWDAF Service Consume subscribes or unsubscribes to the NWDAF by triggering the Nnwdaf_AnalyticsSubscription_Subscribe, or Nnwdaf_AnalyticsSubscription_ Unsubscribe service operation to NWDAF subscribes or unsubscribes to the data analysis results. When the NWDAF receives a subscription-notification, it decides whether to trigger a new data collection.

Step 2: After the NWDAF has received a subscription request, it provides the data analysis results to the NWDAF service consumer via the Nnwdaf_ AnalyticsSubscription_ Notify service operation.

3.6.3.2.1.2 Data analysis results fed to a third-party AF via the NEF

The AF-initiated data analysis result subscription and notification is shown in figure 3-81, where the third-party AF subscribes to the NWDAF via the NEF for data analysis results.

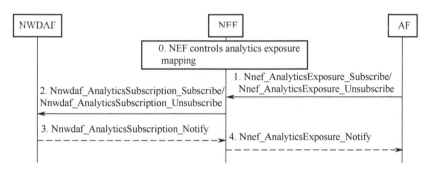

Figure 3-81 Subscription and notification of data analysis results initiated by the AF

The NWDAF feeds the data analysis results to the third-party AF via the NEF, which first authorizes the subscription request service operation, and only allows the subsequent steps to be performed if the authorization is passed. At the same time, the NEF checks the Analytics ID in the AF request based on the operator's policy. Only those that match the operator's policy will trigger a further data analysis result subscription request from the NWDAF.

3.6.3.2.2 Request/response mode

3.6.3.2.2.1 Feedback of data analysis results to the network element

The request and response of data analysis results initiated by the network element is shown in figure 3-82. Based on the request/response model, the consumer of the NWDAF service requests the data analysis results from the NWDAF via Nnwdaf_ AnalyticsInfo service operation.

Step 1: The consumer of the NWDAF service requests data analysis results from the NWDAF by triggering the Nnwdaf_AnalyticsInfo_Request service operation.

Step 2: The NWDAF responds to the data analysis result request via the Nnwdaf_ AnalyticsInfo_Request response service operation.

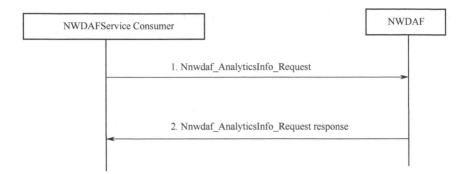

Figure 3-82 Request and response of data analysis results initiated by the network element

3.6.3.2.2.2 Data analysis results fed to third-party AF

The request and response of data analysis results initiated by the AF is shown in figure 3-83, where the third-party AF requests data analysis results from the NWDAF via the NEF.

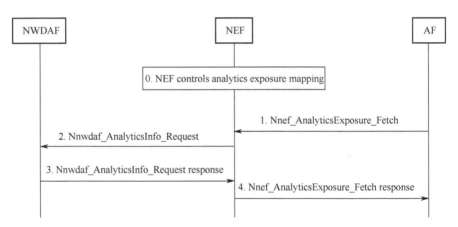

Figure 3-83 Request and response of the data analysis results initiated by the AF

The NWDAF feeds the data analysis results to the third-party AF via the NEF, which first authorizes the subscription request service operation, and only allows the following steps to be performed if the authorization is passed. At the same time, the NEF checks the Analytics ID in the AF request based on the operator's policy, and only those that match the operator's policy will trigger further data analysis result subscription requests from the NWDAF.

3.6.4 Data analysis results available from the NWDAF

This section describes the data analysis results that the NWDAF can provide, the corresponding scenario descriptions, input data, and potential uses, and gives an example of how the NWDAF can obtain data analysis results based on the input data according to an algorithmic implementation. The generic data collection and the feedback process of data analysis results have been described in chapter 2 of this book and will not be repeated in this section.

It should be noted that the input data listed in each case in use in this section are strongly appropriate to the scenario but are not limited to these input data in the specific implementation. Since big data is often uninterpretable and the generalization capability of the model is not permanent, this section only introduces potential input data and service processes, which can be adapted in time to meet service changes during the concrete implementation.

3.6.4.1 Results of service experience data analysis

3.6.4.1.1 Scenario description

The service provider is most interested in the Service Experience (SE) of the service, for example, the Mean Opinion Score (MOS) of the voice service. Compared with the service experience, which can also be measured by the network, the service experience data provided is more accurate because the service provider understands its service logic better. Moreover, the service provider only trusts the service experience data that it has measured itself.

The 5G QoS mechanism introduces a non-standard 5QI that allows operators to tailor the value of QoS parameters to service requirements, thus enabling a better service experience. However, the 3GPP standard has not considered how to set the QoS parameter value in non-standard 5QI to match third-party service experience requirements more effectively, nor how operators can measure the appropriate service experience data, which this section attempts to address based on NWDAF data analysis.

For a specific service, based on the network data as well as the service experience data (obtained from the service provider), the NWDAF can train a service experience model for that service, i.e., the correspondence between the service experience and the network data. Since the service experience in the training data comes from the service provider, the trained service experience model is a good fit for the real service experience model.

If the service provider believes that the wireless network is one of the bottlenecks affecting its service experience, then the service provider should be willing to provide its service experience data to the operator, so that the operator can adjust the wireless network resources to ensure its service quality.

During actual service operation, the NWDAF can measure service experience and adjust network resources more accurately based on network data (real-time or non-real-time), and service experience models. Two examples of this are as follows:

(1) The NWDAF feeds service-level service experience to the PCF, which adjusts the service QoS parameters based on the service experience to ensure the service experience.
(2) The NWDAF feeds back the service experience of the sliced UE to the OAM, assists the OAM in evaluating the sliced Service Level Agreement (SLA), and adjusts the slice resource allocation (e.g., airports, Core Network, transmission network) accordingly.

This section describes how the NWDAF trains the service experience model based on service experience data and network data to measure the service experience and generate service experience data analysis results.

3.6.4.1.2 Input data

The NWDAF needs to collect data from various network elements in order to train the service experience model. The service data collected from the AF is shown in table 3-10; the QoS Flow level network data collected from the 5G Core Network elements is shown in table 3-11, and the UE level network data collected from the network manager is shown in table 3-12.

Table 3-10 Service data from the AF

Types of data (Event ID)	Data source	Description
Application ID	AF	Used to identify a service
Locations of Application	AF/NEF	The area that can be served by a service identified by one or several DNAIs, to which the NEF can be mapped by the AF Service Identifier
Service Experience	AF	Service experience information for the service, such as MOS for voice services, video MOS for video services, or service MOS for the service provider
Timestamp	AF	Corresponding to the time when the AF collects the service experience

Table 3-11 QoS Flow level network data from the 5G Core Network elements

Types of data (Event ID)	Data source	Description
Timestamp	5GC NF	Identifies the time used by network elements to collect data below
Location Info	AMF	Location of UE
DNN	SMF	Identifies the DNN of the PDU session in which the service is located
S-NSSAI	SMF	Identifies the slice in which the service is located
Application ID	PCF/SMF	Service identification provided by the AF to enable the operator network to identify the service type of the QoS Flow
IP filter information	PCF/SMF/UPF	IP filtering information provided by the AF to help the NWDAF identify the service flow
QoS Flow ID (QFI)	PCF/SMF	Identification of the QoS Flow
QoS Flow Bit Rate	UPF	The uplink or downlink bandwidth of the QoS Flow observed by the UPF
QoS Flow Packet Delay	UPF	The uplink or downlink packet delay of the QoS Flow observed by the UPF
Packet retransmission	UPF	The number of packet retransmissions observed by the UPF

Table 3-12 UE level network data from the network manager

Types of data (OAM Service)	Data source	Description
Reference Signal Received Power (RSRP)	OAM	RSRP (including SS-RSRP, CSI-RSRP, E-UTRA RSRP). UE measurements are then reported to the RAN, which further reports to the OAM
Reference Signal Received Quality (RSRQ)	OAM	RSRQ (including SS-RSRQ, CSI-RSRQ, E-UTRA RSRQ). UE measurements are then reported to the RAN, which further reports to the OAM
Signal to Interference plus Noise Ratio (SINR)	OAM	SINR (including SS-SINR, CSI-SINR, E-UTRA SINR). UE measurements are then reported to the RAN, which further reports to the OAM

3.6.4.1.3 Output data analysis results

Based on the service experience model and the network data, the NWDAF can evaluate the service experience of the current service, and the specific service experience information can be described in the following dimensions:

(1) Network data analysis result identifier: Analytics ID = "Service Experience"
(2) Service identification: Application ID
(3) Valid network area for the analysis result: Network area
(4) Valid time window for the analysis result: Time window
(5) Service bandwidth: Media/application bandwidth
(6) Slice identification of the service: S-NSSAI
(7) Name of the data network where the service is located: DNN
(8) Service location: DNAI
(9) List of UE identifiers using the service: List of SUPIs

For clarity, this section also gives an example of how the NWDAF obtains the corresponding service experience data analysis results based on the input data and algorithms.

The NWDAF first correlates the data from each network element based on the correlation identifier to obtain the complete training data. Each sample data in the training data has the following format:

<service experience> <network data, including UE Location,
QoS Flow Bit Rate, etc.>

Based on the above training data and a suitable machine learning algorithm, the NWDAF can train a service experience model, i.e., the Service MOS model. Using linear regression as an example, the Service MOS model can be characterized as:

$$h(x) = w_0 x_0 + w_1 x_1 + w_2 x_2 + w_3 x_3 + w_4 x_4 + w_5 x_5 + \cdots + w_D x_D$$

$h(x)$ denotes the service experience, i.e., Service MOS; x_i $(i = 0, 1, 2, \cdots, D)$ denotes the network data, and D is the dimension of the network data; w_i $(i = 0, 1, 2, \cdots, D)$ is the weight of each network data affecting the service experience, and D is the dimension of the weight.

3.6.4.1.4 Process

The process by which the NWDAF provides service experience data analysis results is shown in figure 3-84.

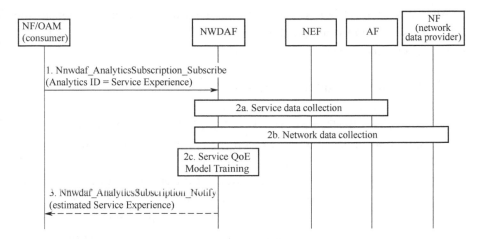

Figure 3-84 The process by which the NWDAF provides service experience data analysis results

Step 1: The NF triggers the service operation of the NWDAF's Nnwdaf_Analyt-
icsSubscription_Subscribe (Analytics ID = Service Experience, Analytics Filter = [Application ID, Time Window, S-NSSAI, DNN, Area of Interest]) for
subscribing to service experience data.

Step 2a: The NWDAF collects the service data on the AF.

Step 2b: The NWDAF collects the network data on the network elements.

Step 2c: The NWDAF trains to obtain a service experience model based on the
service data and network data.

Step 3: The NWDAF is used to indicate the quality-of-service operation by pro-
viding the service experience data analysis results to the NF. Specifically, the
NWDAF can provide data analysis results for two types of Analytics ID =
Service Experience:

(1) Slice-independent: individual service-level (average) service experience

(2) Slice-related: the (average) service experience of individual users, parts of
users (lists or groups of users), and services of all users within a network
slice, where the service level can be one or several services within that
network slice

When the NF is a PCF, the PCF determines whether the service experience
requirements for the current service can be met based on the service experience
information. If not, the PCF adjusts the service QoS parameters, see section 6.1.1.3
and section 6.2.1.2 of 3GPP TS 23.503 [3].

Alternatively, the NWDAF can send service experience information to the OAM to assist the OAM in updating the network slice resources to guarantee the SLA requirements of the slice tenant.

3.6.4.2 Data analysis results of the network element load
3.6.4.2.1 Scenario description

In the Rel-15 phase, the 5G network relies on the NRF for the selection of network elements, such as information on element type, element instance identification, PLMN identification, the slice to which the element belongs, the FQDN or IP address of the element, and the coverage area of the element. However, it does not consider element load information. For example, in an EC scenario where the service provider has very high latency and bandwidth requirements for the service, the SMF needs to refer to the load information of the UPF when selecting the UPF.

In the Rel-16 phase, the eNA topic specifically investigates how the NWDAF generates the network element load based on network data and feeds it to any network element including the NRF, SMF, and AMF to assist in functions such as network element selection.

3.6.4.2.2 Input data

The information that the NWDAF needs to collect in order to obtain network element load information for data analysis is shown in table 3-13.

Table 3-13 Data used for analysis of the network element load information

Types of data (Event ID)	Data source	Description
NF load	NRF	Load information of the network element instance
NF status	NRF	Status of the network element instance
NF resource usage	OAM	Allocation of virtual resources (CPU, memory, and hard disk) of the network element

Further information to be collected by the NWDAF if the load information of the UPF network elements is to be analyzed is shown in table 3-14.

Table 3-14 Data used for analysis of UPF load information

Types of data (Event ID)	Data source	Description
Traffic usage report	UPF	Subscriber's traffic usage report on the UPF

3.6.4.2.3 Output data analysis results

Based on the above data, the NWDAF can obtain two types of network element load information by data analysis:

(1) The statistics for network element load information are shown in table 3-15.

Table 3-15 Network element load information statistics

Types of analysis results	Description
List of resource status (1 ... n)	List of load information observed for each network element instance
> Instance ID	Identification of a network element instance
> NF status	The state of the network element at a given time
> NF resource usage	Average network element resource (CPU, memory, and hard disk) allocation
> NF load	Average load on the network element at a given time
> (Optional) NF peak load	Maximum load on the element at a given time

(2) The network element load information predictions are shown in table 3-16.

Table 3-16 Network element load information predictions

Types of analysis results	Description
List of resource status (1 ... n)	List of load information observed for each network element instance
> Instance ID	Identification of a network element instance
> NF status	The state of the network element at a given time
> NF resource usage	Average network element resource (CPU, memory, and hard disk) allocation
> NF load	Average load on the network element at a given time
> (Optional) NF peak load	Maximum load on the element at a given time
> Confidence	Confidence level of the NWDAF's prediction on the information above

3.6.4.2.4 Process

The process by which the NWDAF provides the analysis results of the network element load data is shown in figure 3-85.

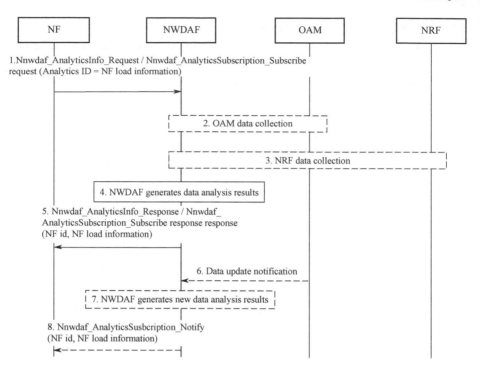

Figure 3-85 The process by which the NWDAF provides analysis results of network element load information

Step 1: The NF triggers the NWDAF's Nnwdaf_AnalyticsInfo_Request or Nnwdaf_ Analytics Subscription_Subscribe request (Analytics ID = NF load information, Analytics Filter = (Time Window), Target of analytics = NF ID) service operation for subscribing to network element load information.

Steps 2 and 3: The NWDAF collects data from the OAM and NRF.

Step 4 to step 8: The NWDAF generates the data analysis results for this network element instance and feeds them back to the NF.

3.6.4.3 Network performance data analysis results

3.6.4.3.1 Scenario description

In the era of big data, the UE side generates huge amounts of data, some of which needs to be uploaded to the Cloud server. These data do not have high requirements for timeliness but are sensitive to the tariffs generated by the transmission traffic, similar to the 3GPP definition of background traffic transmission. For example, there is a strong demand for cars to upload map information to the Cloud server for the development of HD maps. For background traffic transmission, operators can reduce tariffs appropriately to achieve a win-win situation for both the third party and the operator.

In LTE as well as the 5G Rel-15 phase, the AF can provide demand information (including area, number of UEs, and amount of data to be transmitted per UE) to the PCF. The PCF will develop a background traffic transmission policy for the AF based on the network policy, the network load in a particular time and area, and the existing running background traffic transmission policy. However, there are three problems:

(1) How the PCF obtains information on the network load at a given time and in a given area is unclear in terms of standards.
(2) The network condition is constantly changing, and therefore its load is constantly changing, and the background traffic transport policy developed by the PCF may fail over time.
(3) Usually, the network area provided by the AF is large. For example, background traffic transmission policies need to be developed for the whole of Shanghai, and vehicles tend to stay overnight in the neighborhoods where car parks are located, which may lead to network congestion in some neighborhoods if the PCF is developing transmission policies according to the needs of the whole of Shanghai.

The NWDAF can perform data analysis to obtain network performance information (including network load and the number of UEs that can be accommodated) for different time periods within a specific area and provide network performance information within sub-regions (e.g., a cell list) for the network area provided by the AF to assist the PCF in formulating a more accurate background traffic transmission policy.

3.6.4.3.2 Input data
The input data to be collected by the NWDAF in order to obtain network performance data analysis results are shown in table 3-17.

Table 3-17 Input data used for network performance analysis

Types of data	Data source	Description
Load and performance information	OAM	Load and performance information for the RAN corresponding to each cell in the AF demand area
Load information	NRF	Load information for each network element
Number of UEs	AMF	Number of UEs in each cell in the AF demand area

3.6.4.3.3 Output data analysis results

Based on the above data, the NWDAF can obtain two types of network element load information by data analysis:

(1) The network performance information statistics are shown in table 3-18.

Table 3-18 Network performance information statistics

Types of analysis results	Description
Performance information list (1 ... max)	Analysis results of data obtained from statistics in a certain time period
> Network sub-region	TA or Cell ID in the AF demand area
> gNB resource utilization	Statistics on the allocation of gNB resources (CPU, memory, and hard disk) in the network sub-region, including average or maximum values
> Number of UE	Number of UE counted in the network sub-region
> Communication performance	Statistics on the rate of successful PDU session creation in the network sub-region
> Mobility performance	Rate of successful switchover in the network sub-region

(2) The network performance information predictions are shown in table 3-19.

Table 3-19 Network performance information predictions

Types of analysis results	Description
Performance information list (1 ... max)	Analysis results of data obtained from statistics in a certain time period
> Network sub-region	TA or Cell ID in the AF demand area
> gNB resource utilization	Statistics on the allocation of gNB resources (CPU, memory, and hard disk) in the network sub-region, including average or maximum values
> Number of UE	Number of UE counted in the network sub-region
> Communication performance	Statistics on the rate of successful PDU session creation in the network sub-region
> Mobility performance	Rate of successful switchover in the network sub-region
> Confidence	Confidence level of NWDAF predicting the information above

3.6.4.3.4 Process

The NWDAF provides the flow of network performance data analysis results, as shown in figure 3-86.

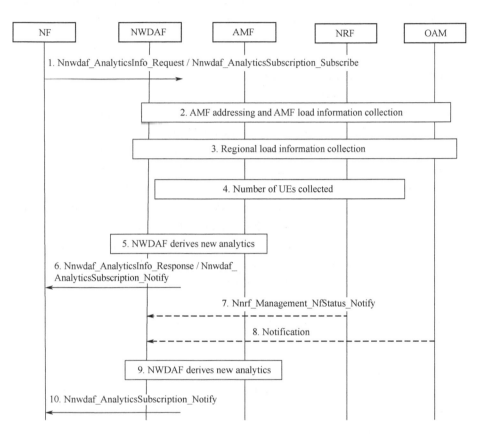

Figure 3-86 The NWDAF provides the network performance data analysis result process

Step 1: NF triggers NWDAF's Nnwdaf_AnalyticsInfo_Request or Nnwdaf_ AnalyticsSubscription_Subscribe (Analytics ID = "Network Performance," Analytics Filter = "Area of Interest" or "Area of Interest" and "Internal Group Identifier," Observation Period (s)) service operation for subscribing or requesting network performance data analysis results.

Step 2: The NWDAF first queries the NRF for information on one or more AMF addresses in the area based on the area information. The NWDAF further queries the NRF for load information of these AMFs.

Step 3: The NWDAF queries the network load information within the area from the OAM.

Step 4: The NWDAF collects the number of UEs from the AMF.

Step 5: The NWDAF generates network performance data analysis results.

Step 6: The NWDAF provides the network performance data analysis results to the NF.

Step 7 to step 10: The NWDAF continuously collects new data from the NRF, OAM, and AMF, then generates data analysis results, and notifies the NF when changes are found in them.

3.6.4.4 UE mobility data analysis results

3.6.4.4.1 Scenario description

In the Rel-15 phase, the concept of the Mobility Pattern (MP) was introduced in 5G networks. The AMF can decide or update the MP based on information such as UE sign-up, local UE location statistics, and operator policies, and then optimize UE mobility-related parameters based on the UE's MP, such as updating the UE's registration area. In addition, as a third party, the AF can also provide the UE's mobility tracking information to the network side to the UDM, and then further provide it to the AMF to assist in the formulation of the MP.

With the introduction of the NWDAF, data analysis can be considered as a way to mine the massive amount of UE location information in order to predict the MP of UE more accurately. For example, the prediction results based on the activity pattern of certain types of people can be fed back to the network side, enabling the AMF to page UE within the cell list, thus reducing the paging burden and saving paging resources.

3.6.4.4.2 Input data

For data analysis to obtain UE mobility data analysis results, the NWDAF can obtain UE mobility information from the OAM, 5G Core Network elements, and the AF on the handset:

(1) The UE location information is collected from the OAM, where the UE location information is present in the MDT data (see 3GPP TS 37.320 [15]).

(2) The UE mobility location data collected from the AMF are shown in table 3-20.

Table 3-20 UE mobility location data collected from the AMF

Types of data	Description
UE identifier	SUPI
UE location list (1 ... max)	
> UE location	The TA or cell that the UE entered
> Timestamp	Time of UE entry into this TA or cell as detected by the AMF
Type Allocation Code (TAC)	UE type identifier indicating the vendor of the terminal. Typically, UEs with the same TAC have similar UE mobility behavior. If some of them do not match this UE mobility behavior, these UEs may have anomalies (see section 3.6.4.6 of this book).
Frequent mobility re-registration	Due to wireless channel quality differences, a stationary UE may re-register between two neighboring cells, which leads to the possibility of a UE registering back and forth between different registration areas. Therefore, the number of frequent mobility re-registrations can indicate whether a UE is anomalous.

(3) The UE mobility trajectory data collected from the AF are shown in table 3-21.

Table 3-21 UE mobility trajectory data collected from the AF

Types of data	Description
UE ID	GPSI or external UE identifier
Service Mark	Service identifier used to identify the service that provides this UE mobility information
UE trajectory (1 ... max)	
> UE location	Location information of the UE
> Timestamp	Time of entry of the UE into the TA or cell

3.6.4.4.3 Output data analysis results

Based on the above data, the NWDAF can analyze and obtain two types of UE mobility data analysis results:

(1) The data analysis results of the UE mobility statistics are shown in table 3-22.

Table 3-22 Data analysis results of the UE mobility statistics

Types of analysis results	Description
UE group identification or UE identification	Identifies a group of UEs (e.g., Internal Group ID as defined in 3GPP TS 23.501 [1]) or a UE (e.g., SUPI)
Time slot information (1 ... max)	
> Start time of the time slot	Start time
> Duration	Duration of the time slot
> List of UE locations (1 ... max)	Observed location statistics
>> UE position	Location information of the UE
>> UE percentage	Ratio of the UEs at this location to the terminal group

(2) The statistical information of the data analysis results of UE mobility trajectory data are shown in table 3-23.

Table 3-23 Statistical information of the data analysis results of the UE mobility trajectory data

Types of analysis results	Description
UE group identification or UE identification	Identifies a group of UEs (e.g., Internal Group ID as defined in 3GPP TS 23.501 [1]) or a UE (e.g., SUPI)
Time slot information (1 ... max)	
> Start time of the time slot	Start time
> Duration	Duration of the time slot
> List of UE locations (1 ... max)	Observed location statistics
>> UE position	Location information of the UE
>> UE percentage	Ratio of UEs at this location to the terminal group
>> Confidence	How much confidence the NWDAF has in predicting the above information

3.6.4.4.4 Process

The NWDAF can provide UE mobility data analysis results to any NF in a statistical form or in a predictive form, as shown in figure 3-87.

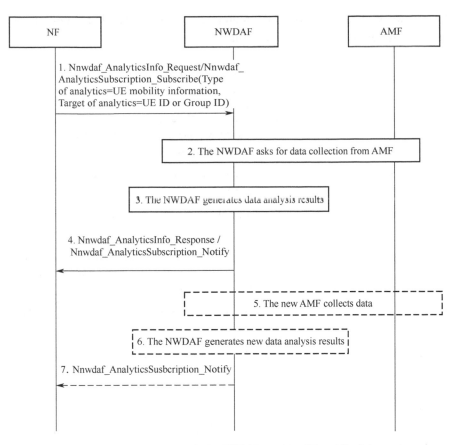

Figure 3-87 The process by which the NWDAF provides UE mobility information

Step 1: The NF requests the UE behavioral data analysis results from the NWDAF via the Nnwdaf_AnalyticsInfo_Request or Nnwdaf_AnalyticsSubscription_ Subscribe service.

Step 2: The NWDAF requests data collection from the AMF.

Step 3: The NWDAF generates the data analysis results.

Step 4: The NWDAF sends the data analysis results to the NF.

Steps 5 to 7: The NF is notified of the updated data analysis results when they change.

3.6.4.5 Results of the UE interactivity analysis

3.6.4.5.1 Scenario description

The third-party AF can provide the UE's Communication Pattern (CP) data to the network side (see table 3-24). The network side can regulate the network behavior of this UE based on this information. For example, the meter in a certain area only reports the utility information of customer power at 9:00 on the Monday of the first week of each month. If the meter sends packets at other times, or sends packets frequently, it may have abnormalities.

Table 3-24 Communication pattern data

Communication pattern parameters	Description
Periodic communication indication	Indicates whether the terminal service data transmission is periodic
Communication duration	Terminal service data transmission duration, e.g., five minutes
Communication period	Terminal service data transmission period, e.g., each hour
Planned communication time	Time zone of terminal service data transmission, such as the day of the week of each week, e.g., time: 13:00–20:00, day: Monday
Static indication	Indicates whether the terminal is stationary or mobile

The above information provided by the third-party AF may not be credible, or the third-party AF may not be able to provide UE interaction pattern data, so the operator may need to generate UE mobility management-related data analysis results directly based on network data.

This section describes how the NWDAF generates data analysis results related to UE mobility management, i.e., UE communication.

3.6.4.5.2 Input data

To generate analysis results for UE communication data, the data to collect are shown in table 3-25.

Table 3-25 UE communication data

Types of data	Data source	Description
UE identifier	AF	GPSI or external UE identification
UE group identifier		Internal Group ID
Link information	AF	Identifies the service of a traffic
Service identifier	AF	Service ID
Communication mode parameters	AF	Same as the Communication Pattern defined in 3GPP TS 23.502 [2]

Types of data	Data source	Description
UE communication information (1 ... max)	AF	
> Communication start time		
> Communication end time		
> Uplink data rate		UE interaction uplink data rate expected by the service (i.e., the number of uplink packets per unit time)
> Downlink data rate		Service expected UE interaction downlink data rate (i.e., the number of downlink packets per unit time)
Type Allocation Code	AMF	UE type identifier indicating the vendor of the UE

Depending on the granularity of the request, or to avoid stressing the network with signaling overhead, the NWDAF can sample from a large number of UEs when collecting data.

3.6.4.5.3 Exporting data analysis results

Based on the above data, the NWDAF can obtain two types of UE communication data analysis results:

(1) The analysis results of the UE communication statistics are shown in table 3-26.

Table 3-26 UE communication statistical analysis results

Types of analysis results	Description
UE group identification or terminal identification	Identifies a group of UEs (e.g., Internal Group ID as defined in 3GPP TS 23.501 [1]) or a UE (e.g., SUPI)
UE communication information (1 ... max)	
> Periodic communication instruction	Indicates whether the UE service data transmission is periodic
> Period	The UE service data transmission period, e.g., per hour
> Planned communication time	The UE service data transmission time zone and day of the week, e.g., time: 13:00–20:00, day: Monday

Types of analysis results	Description
> Start time	Communication start time
> Duration	Communication duration
> Characteristics of flow	DNN, port, etc.
> Method of flow	Uplink or downlink flow (mean or variance)
> Terminal coverage	The percentage of terminals in the UE group under this communication information

(2) Statistical information on the analysis results of the UE communication prediction data is shown in table 3-27.

Table 3-27 Statistical information on the analysis results of the UE communication prediction data

Types of analysis results	Description
UE group identification or terminal identification	Identifies a group of UEs (e.g., Internal Group ID as defined in 3GPP TS 23.501 [1]) or a UE (e.g., SUPI)
UE communication information (1 ... max)	
> Periodic communication instruction	Indicates whether the UE service data transmission is periodic
> Period	The UE service data transmission period, e.g., per hour
> Planned communication time	The UE service data transmission time zone and day of the week, e.g., time: 13:00–20:00, day: Monday
> Time to start	Communication start time
> Duration	Communication duration
> Characteristics of flow	DNN, port, etc.
> Method of flow	Uplink or downlink flow (mean or variance)
> Terminal coverage	The percentage of terminals in the UE group under this communication information
> Confidence	The NWDAF predicts the confidence level of the above information

3.6.4.5.4 Process

The process by which the NWDAF analyzes and provides the analysis results of the UE communication data is shown in figure 3-88.

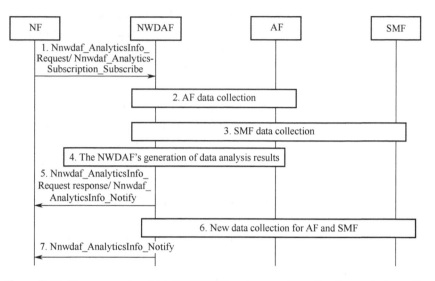

Figure 3-88 The process by which the NWDAF analyzes and provides the analysis results of the UE communication data

Step 1: NF triggers Nnwdaf_AnalyticsInfo_Request or Nnwdaf_Analytics Subscription_Subscribe (Analytics ID = UE communication analytics, Analytics Filter = SUPI) to request or subscribe to data analytics results from the NWDAF.

Step 2 to step 3: The NWDAF collects data from the AF and SMF.

Step 4: The NWDAF generates data analysis results, either statistical UE interaction data analysis results or predictive UE interaction data analysis results.

Step 5: The NWDAF provides data analysis results to the NF.

Step 6 to step 7: When the NWDAF finds that the data analysis results have changed, it notifies the NF about the updated data analysis results.

3.6.4.6 Terminal anomaly analysis results

3.6.4.6.1 Scenario description

The Internet of Things is an important feature of future 5G networks, and a huge number of UEs will be connected to 5G networks in the future. For some specific types of IoT UEs (e.g., smart meters), there are certain patterns in their UE mobility behaviors (e.g., UE movement trajectory) and UE interaction behaviors (e.g., interaction period and interaction duration).

The mIoT UE may be misused or maliciously hijacked for attacking the network, which can lead to serious network security issues such as network congestion, e.g., UEs frequently initiate the registration process of the network due to malicious attacks, causing the UDM to be paralyzed. In addition, IoT UEs and ordinary UEs have the possibility of being attacked or hijacked, so 5G networks need a mechanism to enable the regulation of UE behavior. With the help of big data analytics, the NWDAF can perform IoT UE monitoring and management.

This section explains two methods by which the NWDAF can analyze whether a UE is abnormal or not.

The first is the unsupervised learning method. In off-line analysis, the NWDAF collects UE behavior data in normal network conditions and categorizes these UE with an unsupervised classification algorithm. Each class of UE has common UE behavior characteristics (i.e., expected UE behavior data), and pushes it to the UDM as the contracted data of this class of UE; in online detection, if the real-time UE behavior data does not match the UE behavior characteristics, the UE is considered abnormal.

The second is the supervised learning method. In off-line analysis, the NWDAF obtains UE anomaly-type labels from firewalls or shared security intelligence centers, and then collects the corresponding UE behavior data from the network side so that the NWDAF can train to obtain unexpected UE classification models based on classification algorithms (e.g., support vector machines). During online detection, the real-time behavior data of the UE is input to this abnormal UE classification model to determine whether the UE is abnormal and the specific abnormal type.

It is important to note that firewalls or shared security intelligence centers are both application-layer detections; that is, UE data packets have passed through the network and have a negative impact on the network. This section mainly focuses on abnormal monitoring and analysis of UE behavior at the network layer, which can prevent possible network attacks more effectively. In addition, the solution in this section also supports the detection of non-application layer attacks, such as UE network location anomaly, UE session time anomaly, and pseudo base station detection.

For UEs that do not conform to user behavior rules, the NWDAF needs to promptly notify PCF policy network elements or SMF user management network elements to formulate corresponding processing strategies, for example, denying abnormal UEs access in order to avoid network congestion and guarantee normal network operation.

3.6.4.6.2 Input data

The input data that the NWDAF needs to collect to implement UE anomaly detection are as follows:

(1) Real-time UE behavior data, including UE mobility data described in section 3.6.4.4 of this book and the UE communication data described in section 3.6.4.5.
(2) Expected network-side UE behavior data, including the results of the analysis of UE mobility data described in section 3.6.4.4 of this book, and the results of the analysis of UE communication data described in section 3.6.4.5.
(3) From the security designer's perspective, a firewall or threat-sharing intelligence center (treated as a third-party AF) can provide anomalous types of a certain service flow (identified by the IP 5-tuple). See table 3-28 for abnormal UE behavior data collected from the AF.

Table 3-28 Abnormal UE behavior data collected from the AF

Types of data	Description
IP 5-tuple	Identifies a service flow
Unexpected Information (1 … max)	
> Identifier of an anomaly	Identifies an unexpected type, e.g., extra-long traffic, extra-large traffic, Distributed Denial of Service (DDoS) attack, Advanced Persistent Threat (APT) attack.
> Level of the anomaly	Indicates the severity of the unexpected type
> Tendency of the anomaly	Indicating the type of anomaly (up, down, smooth, and unknown)

3.6.4.6.3 Output data analysis results

The analysis results of the specific unexpected UE behavior data output by the NWDAF are shown in table 3-29.

Table 3-29 Analysis results of unexpected UE behavior data

Types of analysis results	Description
UE identifier	Can be SUPI, inner UE group identifier, external UE identifier, or TAC.
Unexpected Information (1 … max)	
> Identifier of an anomaly	Identifies an unexpected type, such as extra-long traffic, extra-large traffic, DDoS attacks, and APT attacks
> Level of the anomaly	Indicates the severity of the unexpected type
> Tendency of the anomaly	Indicates the trend of the unexpected type (up, down, smooth, and unknown)
> Identifier of the anomaly	Identifies an unexpected type, such as extra-long traffic, extra-large traffic, DDoS attacks, and APT attacks

Different types of unexpected UE behavior and the corresponding coping strategies are shown in table 3-30.

Table 3-30 Types of unexpected UE behavior and corresponding coping strategies

Types of unexpected behavior (Exception ID)	Policy network element	Coping strategies
Unexpected long-live/ large rate flows	SMF	If it is not dynamic PCC, the SMF updates the PCC rules to reduce stream bandwidth.
	PCF	If it is dynamic PCC, the SMF updates the PCC rules to reduce the flow bandwidth.
Suspicion of a DDoS attack	SMF	If it is not dynamic PCC, the SMF updates the packet filter to block the packets sent to the destination address corresponding to DDoS.
	PCF	If it is dynamic PCC, the PCF updates the packet filter to block the packets sent to the destination address corresponding to DDoS.
Wrong destination address	PCF	The PCF updates the PCC Rule and notifies the UPF via the SMF to discard uplink and downlink packets from this address.
Unexpected UE location	AMF	Updates the mobile restriction area
	PCF	Updates the mobile restricted area

Types of unexpected behavior (Exception ID)	Policy network element	Coping strategies
Frequent service access or unexpected flow	AF	The AF blocks service access or abnormal traffic at the application layer.
Unexpected wake up	AMF	The AMF updates the MM back-off timer.
Ping-pong stationary UE	AMF	Updates the mobile restricted area

The process by which the NWDAF analyzes the unexpected behavior of a UE is shown in figure 3-89.

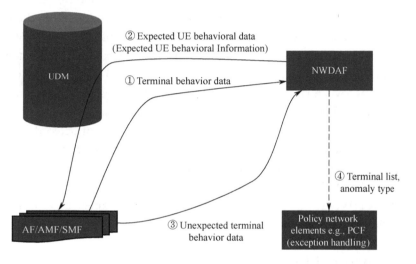

Figure 3-89 The process by which the NWDAF analyzes the unexpected behavior of a UE

Step 1: Data collection and pre-processing. The NWDAF collects a large amount of UE behavioral data from the network side, including

(1) UE mobility data: UE interactivity data, which includes normal and abnormal types of UEs;

(2) abnormal UE data, which includes only abnormal types of UE.

Based on the IP 5-tuple, the NWDAF correlates the abnormal UE data collected from the AF with the communication data collected from the network side. The training data set can be divided into two parts: normal UE behavior data and abnormal UE behavior data.

Step 2: The UE behavior data expected by the network side is distributed. The NWDAF performs clustering based on normal UE behavior data, for example, using the K-Means clustering algorithm. In the clustering result, the center of mass of each category cluster is the network-side expected UE behavior data of a group of UEs in that category.

The network-side expected UE behavior data includes two categories of UE mobility and UE communication, the former will be distributed to AMF for UE mobility behavior monitoring, and the latter will be distributed to SMF for UE communication behavior monitoring. There are three main ways:

(1) The NWDAF first sends the expected UE behavior data from the network side to the AF, which then sends it to the UDM via the NEF; the UDM stores it as UE contract data, (see section 4.15.6 of 3GPP TS 23.502 [2]) for the process. Eventually, the UE behavior data expected by the network side is distributed to the AMF as signing data and to the SMF as UE context storage when UE registers and when UE establishes a PDU session.

(2) The NWDAF sends the expected UE behavior data from the network side to the UDM, and the UDM stores it as the contracted data of the UE. Finally, when UE registers and when it establishes a PDU session, the expected UE behavior data from the network side is distributed to the AMF as UE context storage and to the SMF as the contracted data.

(3) The NWDAF directly sends the UE behavior data expected by the network side to the AMF or SMF.

Step 3: Unexpected UE behavior data reporting. If the NWDAF wants to detect whether the UE is abnormal in real time, it needs to collect further UE behavior data from the AMF and SMF.

In order to avoid potential signaling storms being brought to the network by data reporting, the AMF or SMF can perform advanced detection, that is, to compare (based on a certain threshold) the real-time UE behavior data with the expected UE behavior data on the network side, obtained from the UDM or NWDAF based on step 2 above. When a mismatch is found, the AMF or SMF considers that the UE may have anomalies but does not know the UE abnormality type. Then, the AMF or SMF will further report this real-time UE behavior data.

Step 4: The NWDAF analyzes the UE anomaly and notifies the policy network element PCF. For clarity, this section also gives an example of how the NWDAF gains the corresponding service experience data analysis results based on certain input data and a certain algorithm.

First, based on the training data obtained from step 1 (including normal UE behavior data and abnormal UE behavior data), supervised machine learning algorithms, such as Logistic Regression (LR) and Support Vector Machine (SVM), are used to train the classifier. Using LR as an example, the classifier (normal type and abnormal type in the case of binary classification) can be represented as

$$y_i = \begin{cases} 0, & z_i < 0 \\ 1, & z_i \geqslant 0 \end{cases}$$

among which,

$$z_i = w_0 \cdot x_{i0} + w_1 \cdot x_{i1} + w_2 \cdot x_{i2} + w_3 \cdot x_{i3} + \cdots + w_D \cdot x_{iD} \tag{1}$$

In the above two equations:

(1) y_i is the classification result of the behavioral data of the ith UE; if $y_i = 1$, then the UE behavioral data is abnormal; if $y_i = 0$, then the UE behavioral data is normal;

(2) z_i is the intermediate data value of x_i obtained after linear regression;

(3) $x_i = \{x_{i0}, x_{i1}, x_{i2}, x_{i3}, \cdots, x_{iD}\}$ is the vector converted from the behavioral data of the ith UE, in which $x_{i0}, x_{i1}, x_{i2}, x_{i3}, \cdots, x_{iD}$ is the behavioral data of the terminal, such as the start time of communication or interaction, uplink or downlink packet delay, and the number of frequent mobile re-registration (see table 3-20, table 3-21 and table 3-25);

(4) $w = \{w_0, w_1, w_2, w_3, \ldots, w_D\}$ is for weights.

In real-time detection, the NWDAF first converts the real-time UE behavior data into a vector and then inputs it to this classifier, i.e., equation (1), and determines the UE to be of normal type, if the value obtained is 0, and of abnormal type if the value obtained is 1.

The above algorithm example is binary classification. If it is multi-classification, such as the training data set in table 3-30, a classifier with four classifications (normal type, abnormal type X, abnormal type Y, and abnormal type Z) is needed. There are many multi-classification machine learning algorithms, but they will not be discussed here.

The NWDAF then sends the UE identification and the corresponding UE's abnormal type to the policy network element, which develops the abnormal UE processing policy and informs the network side to execute it.

3.6.4.6.4 Process
The process by which the NWDAF analyzes and provides the analysis results of unexpected UE behavior data is shown in figure 3-90.

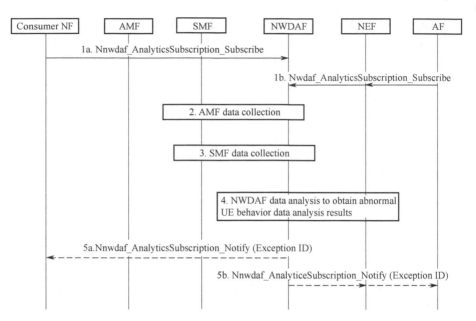

Figure 3-90 The process by which the NWDAF analyzes and provides the analysis results of unexpected UE behavior data

Step 1: The NF or AF triggers the Nnwdaf_AnalyticsSubscription_Subscribe (Analytics ID, Analytics Filter (s) = Internal-Group-Identifier or SUPI or External-Group-Identifier or External UE ID) service of the NWDAF to subscribe to the UE's abnormal data analysis results.

Step 2 to step 3: The NWDAF collects UE behavior data. The AMF or SMF needs to perform local detection of UEs for abnormalities before reporting UE behavior data.

Step 4: The NWDAF performs data analysis to identify hijacked or misused UEs and gets the specific anomaly type of that UE.

Step 5: The NWDAF notifies the NF or AF of the corresponding UE identification and the anomaly type. Information on the collection of normal UE behavior by the NF and AF can be found in section 3.6.4.6.3 of this book.

3.6.4.7 Other data analysis results

In addition to the above data analysis results, 3GPP also provides data analysis results such as user data congestion and QoS prediction.

1. ANALYSIS RESULTS OF USER DATA CONGESTION INFORMATION

The NWDAF can provide user data congestion information. User data can be transmitted through the control plane or the user plane. The granularity of user data congestion information can be either of area granularity or of UE granularity. Furthermore, the NWDAF can provide one-time or continuous user data congestion information to the consumer network elements of its analysis result.

The consumer network elements of the analysis result can be a third-party AF element. The AF can assist the application layer in making decisions, such as adjusting the cache window size, after obtaining the user-side data congestion information.

2. QOS PREDICTION DATA ANALYSIS RESULTS

V2X's demand for QoS prediction is shown in figure 3-91. The remote driving or autonomous driving remote/autonomous driving in V2X requires a high-quality network. For example, the end-to-end delay should be below 5 ms, and the reliability should reach 99.999%. In the early stages of 5G deployment, there are weak coverage areas for V2X services, requiring the carrier's network to be able to notify the AF of network QoS quality changes in the vehicle's path of travel.

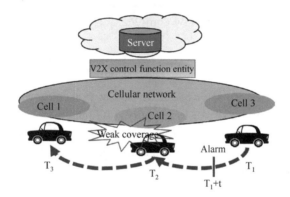

Figure 3-91 V2X's demand for QoS prediction

The NWDAF can assist in analyzing potential V2X service QoS quality changes on the user path, and promptly notify V2X servers to adjust application layer decisions based on network changes, such as network driving levels.

3.6.5 Summary

Section 3.6 of this book focuses on the Rel-15 and Rel-16 parts of the 3GPP SA2 work on the NWDAF.

The Rel-15 5G network architecture introduces the NWDAF and defines the analysis results of the network element load data. Rel-16 E-UTRANC (eNA) Issue further defines smart 5G network architecture and opens up the control plane, management plane, and application servers to enable data collection. At the same time, relevant data analysis results such as service experience data analysis results, terminal mobility data analysis results, and UE anomaly analysis results were defined.

3.7 IoT

3.7.1 Introduction

Massive Machine Type Communications (mMTC) is one of the main scenarios for the supply of 5G networks, mainly for various types of Internet of Things (IoT) service based on cellular networks. The mMTC application scenarios are characterized by the connectivity needs of massive IoT terminals, which can reach 1 million connections per square kilometer, posing a huge challenge to its network architecture and protocol design. In addition, mMTC applications have features including low data volume (small packet transmission of tens of bytes), low power consumption (the battery life of IoT terminals can reach 10 years), enhanced coverage (supporting various weak coverage scenarios, such as in underground garages or elevator shafts), and low complexity (reducing terminal and network costs).

3.7.2 NB-IoT

To meet the needs of large-scale connectivity, low power consumption, enhanced coverage, and low complexity of IoT services, 4G networks have been optimized accordingly, defining NB-IoT (Narrow Band Internet of Things) wireless access technology and introducing standard features such as overload control, routing, small data transmission, terminal power-saving mode, and QoS enhancement, as shown in table 3-31.

Table 3-31 Main characteristics of 4G IoT

Standard version	Main characteristics of IoT
Release 10	(1) MME/PGW congestion control, SGW side downlink paging suppression (2) RRC connection request rejection, Extended Access Barring (EAB) (3) Long-periodic location update timer
Release 11	(1) IPv6-based routing addressing for IoT terminals (2) PS only service (3) MTC Device Trigger
Release 12	(1) Small data transmission optimization: Core Network-assisted eNB parameter adjustment (2) MTC device power consumption optimization: Power Saving Mode (PSM)

Standard version	Main characteristics of IoT
Release 13	(1) NB-IoT and eMTC (2) Small data transmission: the control plane optimization scheme (NAS signaling to transmit small packets), the user plane optimization scheme (connection hang/resume). (3) Low power optimization: extended DRX (eDRX), high-latency communication. (4) Dedicated Core Network (DCN) (5) Open network capacity (6) Cluster optimization
Release 14	(1) Coverage enhancement authorization (2) Location, broadcast multicast service enhancement (3) Overload control and QoS enhancement for small data transmission on the control plane (4) Inter-RAT idle state mobility enhancement
Release 15	(1) Service Gap control (2) Core Network eMTC traffic recognition (3) Early Data Transmission (EDT)

To support the access of NB-IoT terminals to the 5G Core Network, the 5G network architecture migrates and adapts the IoT features of existing 4G networks and minimizes the impact on existing NB-IoT terminals to achieve smooth evolution. 5G networks support the following IoT features:

(1) small data transmission
(2) terminal power-saving functions
(3) high-latency communication
(4) coverage enhancement management
(5) small data overload control
(6) non-IP data reliable transmission services
(7) EPC-5GC interoperable unified northbound API interface
(8) API interface-based configuration of network parameters and UE behavior parameters
(9) event monitoring
(10) idle-state Inter-RAT mobility for NB-IoT
(11) NB-IoT QoS enhancement
(12) Core Network node selection and redirection
(13) unicast-based group messaging
(14) SMS transmission without MSISDN (MSISDN-less Mobile Originated [MO] SMS)

3.7.3 Key technological solutions for 5G to support IoT

3.7.3.1 Small data transmission

The conventional data transmission process involves signaling interactions such as PDU session management and user-plane bearer establishment, which causes the signaling overhead caused by transmitting small data packets to far exceed the amount of data to be transmitted. Therefore, small data transmission mainly addresses how to simplify the complex signaling interactions required for data transmission to improve transmission efficiency.

There are two optimization schemes for small data transmission, namely the control plane optimization scheme and the user plane optimization scheme.

The UE and AMF negotiate the supported and used small data transmission scheme through the registration process. UE carries 5G Preferred Network Behavior cells in the registration request, which points to the 5G IoT optimization scheme supported and preferred by the terminal. They include the following information:

(1)   whether Control Plane C-IoT 5GS Optimization is supported
(2)   whether User Plane C-IoT 5GS Optimization is supported
(3)   whether the control plane optimization scheme or the user plane optimization scheme is preferred
(4)   whether regular N3 user plane data transmission is supported
(5)   whether Control Plane Optimization header compression is supported

AMF determines the small data transmission scheme supported by the UE based on information such as local configuration and carrier policy and returns the result to the UE carrying the 5G Supported Network Behavior cells in the registration acceptance request. The result includes the following information:

(1)   whether the control plane optimization scheme is supported
(2)   whether the user plane optimization scheme is supported
(3)   whether regular N3 user plane data transmission is supported
(4)   whether Control Plane Optimization header compression is supported

1. CONTROL PLANE OPTIMIZATION SCHEME

The control plane optimization scheme optimizes small data transmission by encapsulating IP data, Ethernet data, and non-IP data (also called unstructured data) into NAS-SM Protocol Data Units (PDUs) without establishing user plane connections.

Based on the existing NAS PDU encryption and integrity protection mechanisms, the UE and AMF encrypt and protect the integrity of user data.

For IP and Ethernet data, the UE and SMF perform header compression based on the Robust Header Compression (ROHC) framework defined by IETF RFC 5795 and determine ROHC configuration and ROHC context establishment through PDU session establishment requests.

The uplink for the small data transmission process of the Control Plane Optimization scheme is shown in figure 3-92. The downlink for small data transmission process is similar to the uplink transmission process and is hence not further described.

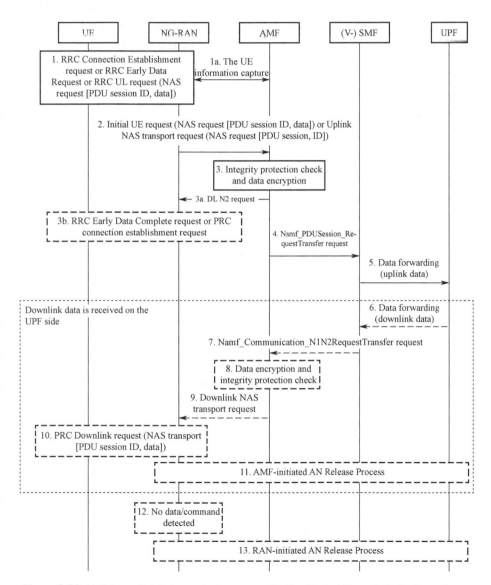

Figure 3-92 Uplink small data transmission process for the Control Plane Optimization scheme

Step 1: If the UE is in the connected state, it sends an NAS request carrying the PDU session ID and small data to the AMF via the NG-RAN. If the UE is idle, it needs to send an RRC Connection Establishment request to the NG-RAN, and then send an NAS request with data after the RRC connection is established or send an NAS request with data via an RRC Early Data Request. The UE can also send the AS Release Assistance Information (AS RAI) to the NG-RAN in the RRC request or carry the NAS Release Assistance Information (NAS RAI) in the NAS request. NAS RAI is used to indicate whether there are still uplink or downlink packets to be sent after the uplink packet transmission, or whether there is only one downlink packet to be sent (e.g., the ACK or response corresponding to the UL packet).

Step 1a: For NB-IoT access, the NG-RAN can obtain NB-IoT UE Priority and Expected UE Behavior from the AMF. NB-IoT UE Priority is used by the NG-RAN to distinguish the priority of different UEs during the RRC connection establishment process or before triggering step 2. The NG-RAN can also obtain other required parameters from the AMF, such as UE Capability Information.

Step 2: NG-RAN transports NAS PDUs to the AMF via the N2 Initial UE request and carries the "EDT Session command in the N2 Initial UE request if the RRC Early Data Request is received in step 1. If the AS RAI command received in step 1 indicates that there is no subsequent uplink or downlink data to be sent, the NG-RAN carries the RRC Connection Release Request in the N2 Initial UE request.

Step 3: The AMF checks the integrity of the NAS request, and then decrypts the data.

Step 3a: If the EDT Session command from the NG-RAN is received in step 2, the AMF sends an N2 request to the NG-RAN.

(1) If the NAS RAI indicates that there is no downlink data to be sent, or if an RRC connection release request is received from the NG-RAN and there is no downlink data or command to be sent on the AMF side, the AMF carries End Indication in the N2 request to indicate that there is no subsequent data or command to be sent by the UE.

(2) If the AMF detects pending data or commands, the AMF carries the N2 connection establishment indication message in the N2 request.

Step 3b: If step 3a is executed, the NG-RAN completes the RRC EDT process according to the following processes:

(1) If this is a step 3a (1) scenario, the NG-RAN sends an RRC Early Data Complete message to the UE. The entire uplink data transfer process ends after step 5.

(2) If this is a step 3a (2) scenario, the NG-RAN initiates the RRC connection establishment process. The entire uplink data transfer process ends after the completion of step 12.

Step 4: The AMF determines the corresponding (V-)SMF based on the PDU session ID in the NAS request and sends the decrypted data and the PDU session ID to the (V-)SMF.

Step 5: If header compression is enabled, the (V-)SMF needs to decompress the received data. The V-SMF further forward the data to the H-SMF in a roaming scenario. The SMF then transports the data to the UPF, which sends the data to the corresponding data network according to the routing forwarding rules.

Step 6: If the UPF receives the downlink data, the UPF forward the data to the V-SMF. The H-SMF further forward the data to the V-SMF in a roaming scenario.

Step 7: If header compression is enabled, the V-SMF needs to perform header compression on the received data. The V-SMF sends the downlink data and the corresponding PDU session ID to the AMF.

Step 8: The AMF creates a Downlink NAS transport request with downlink data and PDU session ID. The AMF further encrypts the message and performs integrity checks.

Step 9: The AMF sends a Downlink NAS transport request to the NG-RAN.

Step 10: The NG-RAN sends the NAS request down to the UE via the RRC request.

Step 11: If the AMF has no data or command to be sent and the RAI sent by the UE indicates that there is no subsequent downlink data, the AMF triggers the AN Release Process.

Step 12: If no signaling/data transmission is detected for a period of time, NG-RAN triggers the AN Release Process.

Step 13: The AN Release Process is performed, and the terminal releases the signaling connection and completes data transmission.

2. User plane optimization scheme

The user plane optimization scheme optimizes the user plan connection rebuilding process. The NG-RAN keeps the context information when the UE turns idle, thus eliminating the need to use the Service Request process to re-establish the user plane bearer and security context between the NG-RAN and the UE when the UE transmits data in the Idle state. To maintain the movement of the UE between different NG-RANs, the UE context information can be forwarded between NG-RANs.

The user plane optimization scheme mainly consists of the Connection Suspend and Connection Resume processes. When NG-RAN triggers the Connection Suspend process, the UE, NG-RAN, and AMF save the UE's context information. When the UE triggers the Connection Resume process, the UE uses the stored AS context information to restore the NG-RAN connection; the NG-RAN informs the AMF of the UE connection restoration information, and the AMF triggers the SMF to activate the user plane resources corresponding to the UE's PDU session.

Figure 3-93 shows the connection suspend process of the user plane optimization scheme.

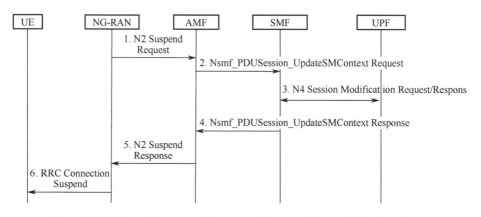

Figure 3-93 Connection suspend process of the user plane optimization scheme

Step 1: The NG-RAN sends an N2 Suspend Request message to the AMF to trigger the Connection Suspend process. The AMF enters the CM-IDLE state and records that the UE's connection has been suspended. The NG-RAN, UE, and AMF maintain the NGAP UE association, the UE context, and PDU session context required to restore the connection. The NG-RAN may also include in the N2 Suspend Request message recommended cell, and the NG-RAN information and coverage enhancement information for subsequent paging.

Step 2: For the suspended PDU session, the AMF sends the Nsmf_PDUSession_ UpdateSMContext Request message to the corresponding SMF, instructing the SMF to suspend the user plane resources corresponding to the PDU session.

Step 3: The SMF sends an N4 Session Modification Request message to the UPF, instructing the latter to release the corresponding NG-RAN user plane tunnel information and whether caching is required when downlink data is received. The UPF returns the execution result to the SMF.

Step 4: The SMF sends the Nsmf_PDUSession_UpdateSMContext Response message to the AMF to return the execution result.

Step 5: The AMF sends an N2 Suspend Response message to the NG-RAN to successfully terminate the Connection Suspend process initiated by the NG-RAN.

Step 6: The NG-RAN sends an RRC message to the UE to hook up the RRC connection to the UE.

Figure 3-94 shows the connection resume process of the user plane optimization scheme.

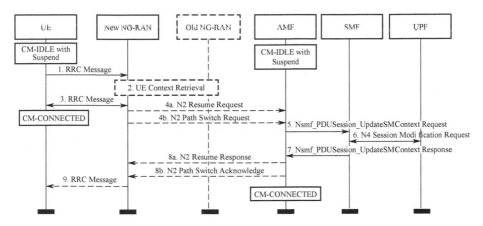

Figure 3-94 Connection resume process for the user plane optimization scheme

Step 1: The UE sends an RRC request to trigger the connection recovery process and provides the Resume ID in the RRC message for the NG-RAN to query the stored UE context information.

Step 2: If the UE is connected to the new NG-RAN, the new NG-RAN tries to get the context information of the UE from the old NG-RAN.

Step 3: If the NG-RAN successfully obtains the UE context information, the NG-RAN and the UE synchronize the access layer configuration information via RRC messages.

Step 4a: If the UE does not alter the NG-RAN, the NG-RAN sends an N2 Resume Request message to the AMF to notify the UE of the RRC connection restoration, which carries the N2 SM information to indicate the successfully restored and unsuccessfully restored PDU sessions.

Step 4b: If the UE alters the NG-RAN and the new NG-RAN obtains the UE's context information from the old NG-RAN, the new NG-RAN sends an N2 Path Switch Request message to the AMF to notify the UE of the RRC connection restoration, which carries the NG-RAN tunnel information and indicates the successfully switched and unsuccessfully switched PDU sessions.

Step 5: For each successfully resumed PDU session indicated in step 4, the AMF sends a Nsmf_PDUSession_UpdateSMContext Request to the corresponding SMF instructing the SMF to restore the user plane resources of the corresponding PDU session.

Step 6: The SMF sends an N4 Session Modification Request to the UPFm indicating the AN tunnel information corresponding to the PDU session that needs to be recovered and whether caching is required when downlink data is received. The UPF sends an acknowledgment message back to the SMF.

Step 7: The SMF sends the Nsmf_PDUSession_UpdateSMContext Response message to the AMF, which may include new CN tunnel information (if the SMF assigns new CN tunnel information for the recovered PDU session) as well as information about the PDU session that has failed to recover.

Step 8: The AMF sends an N2 Resume Response message (in response to the message received in step 4a) or an N2 Path Switch Acknowledge (in response to the message received in step 4b) to the NG-RAN, indicating either the resumed PDU session information or the failure of the recovery.

Step 9: The NG-RAN sends an RRC message to the UE indicating the result of the connection restoration.

3. Non-IP Data Delivery (NIDD) scheme

The functions of Non-IP Data (also called Unstructured Data) include the MO and the Mobile Terminated (MT) transmission, and support the following two transfer methods:

(1)   Non-IP data transmission based on the NIDD API provided by NEF
(2)   Non-IP data transmission based on UPF point-to-point N6 tunnel

During the PDU session establishment process, the SMF determines whether the non-IP data transmission needs to use the NIDD API transmission method based on the UE contract information. If the UE contract information contains the DNN/S-NSSAI related "NEF Identity for NIDD" information, the SMF selects the corresponding NEF for data transmission based on the NIDD API.

The NIDD API-based uplink data transmission process is shown in figure 3-95. The downlink transmission process is similar to the uplink and is hence omitted in the description.

Step 1: According to the control plane optimization scheme, the UE sends non-IP data to the AMF via NAS messages, which are forwarded by the AMF to the corresponding SMF (see steps 1 to 4 of the control plane optimization scheme uplink data transmission process for details).

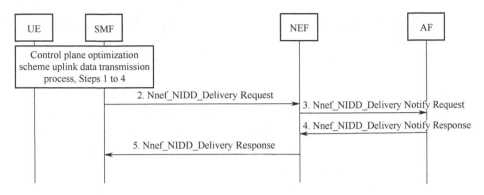

Figure 3-95 Uplink data transmission process based on NIDD API

Step 2: The SMF sends a Nnef_NIDD_Delivery Request message to the NEF containing the UE identity, NEF ID, and non-IP data.

Step 3: The NEF obtains the corresponding target AF address based on the stored NEF PDU session context and sends the non-IP data to the target AF via Nnef_NIDD_Delivery Notify Request. If the NEF does not obtain the corresponding target AF address, the non-IP data will be discarded.

Step 4: The AF sends a Nnef_NIDD_Delivery Notify Response (reason value) message to the NEF indicating whether the data has been received.

Step 5: The NEF sends a Nnef_NIDD_Delivery Response message to SMF to return the data delivery result.

3.7.3.2 Terminal power-saving function

As IoT terminals are required to operate on lower levels of energy consumption, 5G systems have the following two types of terminal power-saving functions:

(1) They support extended DRX (eDRX) function, including the eDRX function of the terminal idle state, and the eDRX function of the terminal in RRC-Inactive state.

(2) They support MICO mode enhancement, including Active Time negotiation, Connected Time enhancement, and MICO mode control based on Expected UC Behavior.

1. EDRX FUNCTIONS

The terminal idle-state eDRX mechanism reuses the EPC idle-state eDRX mechanism as follows:

(1) eDRX period negotiation: The UE negotiates the eDRX period with AMF in the NAS registration process. AMF determines the final use of the eDRX

period according to the UE contract, local configuration policy, and other information.

(2) Signing information: The UE signing information in UDR increases the eDRX cycle value of UE.

(3) H-SFN (Hyper-System Frame Number), PH (Paging H-SFN), and PTW (Paging Time Window) processing: adopt similar processing as EPC and assigns PTW with AMF. The eDRX cycle and PTW parameters are sent by AMF to the UE via NAS messages and to the NG-RAN via paging message, to enable paging by NG-RAN.

(4) For eMTC, the NG-RAN needs to broadcast in the system message whether eDRX is supported.

Below are the functions of eDRX under CM-CONNECTED/RRC-Inactive status and the functions of the RAN to cache downlink data and handle the associated paging:

(1) Similar to the CM-Idle state, the UE requests the idle state eDRX parameter from the AMF via a NAS message.

(2) The AMF sends the finalized idle-state eDRX parameters to the NG-RAN.

(3) If the UE supports eDRX in RRC Inactive function, the NG-RAN determines the eDRX parameter of RRC-Inactive state based on the idle state eDRX parameter and the maximum value of NAS/SMS retransmission timer. In general, the eDRX period of the RRC-Inactive state is much smaller than the eDRX period in the idle state.

(4) NG-RAN caches downlink data during the eDRX cycle and processes paging according to the eDRX parameters of the RRC-Inactive state.

2. MICO MODEL ENHANCEMENT

For the MICO mode already supported by 5GS, an Active timer parameter similar to PSM is further introduced, i.e., the UE and AMF can negotiate whether to adopt MICO mode and the corresponding Active timer parameter value via NAS messages.

In the existing MICO mode, when the data transmission ends and goes to the idle state, the terminal immediately goes to sleep mode and cannot receive the subsequent arriving downlink data in time. 5G IoT introduces Connection Time to solve this problem, as is illustrated as follows:

(1) When the AMF determines that there will be subsequent MT data, or that MT SMS, network-initiated signaling, or other downlink messages are to be

sent, the AMF maintains the N2 connection and provides the connection time parameters to the NG-RAN.

(2) The connection time parameter is used to indicate the minimum time that the NG-RAN shall keep the UE in RRC_ CONNECTED. The AMF may determine the connection time parameter based on local configuration information, maximum response time provided by the UDM, and other information.

(3) If the NG-RAN node receives the connection time parameter value from the AMF, the NG-RAN keeps the UE in the RRC_CONNECTED state according to the connection time parameter value to receive subsequent MT data or signaling messages.

(4) When the connection time parameter timer times out, the UE starts the inactivation timer. When the inactivation timer times out, the NG-RAN releases the RRC connection of the UE.

Considering that parameters such as the desired behavior of the UE may affect the configuration of MICO mode, 5G IoT supports MICO mode management based on the desired behavior of the UE. If the UE instructs the AMF to request MICO mode in a NAS message and the AMF obtains information on the desired behavior parameters of the UE from the UDM, the AMF can enhance the MICO management.

If the value of the desired behavior parameter of the UE indicates that the downlink communication of the UE is performed during a certain period, the AMF will set the UE to MICO mode between the respective downlink communication scheduling periods. For example, if the downlink communication scheduling period of the UE is 13:00 to 14:00 every day, the UE may be in MICO mode at times other than that period.

Also, the AMF can assign the appropriate periodic registration update timer to the UE so that the UE performs the periodic registration update process before or within the downlink communication scheduling period, thus preventing the UE from waking up from MICO mode to perform the periodic registration update process at other periods.

In addition, to prevent the UE's uplink transmission from breaking the synchronization between the UE and the AMF (downlink communication scheduling time slot synchronization), the AMF can also send a "do not reset the timer for Periodic Registration" message to the UE so that the UE can regularly perform the periodic registration update.

3.7.3.3 High-latency communication

High-latency communication is mainly for terminals with power-saving functions (e.g., MICO, eDRX). The network reduces network congestion and resources waste by efficiently handling downlink data for delay-tolerant services and avoiding unnecessary data transmission.

High-latency communication in 5G IoT basically reuses the EPC scheme, mainly including the extended caching scheme and event notification scheme.

1. EXTENDED CACHING SCHEME

When downlink data is received and if the UE adopts the power-saving function resulting in unreachability, the AMF instructs the SMF to cache the downlink data and wait until the UE transitions to the connected state before distributing the cached data to the terminal. The data caching point can be at the SMF or UPF. The specific process includes when downlink data is received from the UE, the SMF determines whether to enable the extended caching function according to the policy configuration and instructs the AMF. The AMF receives the downlink data notification from the SMF and the cache function indication information. If the UE adopts the power-saving function and results in unreachability, the AMF instructs the SMF to cache the data and provides the maximum waiting time to the SMF according to the parameter setting related to the power-saving function. The SMF caches the downlink data or instructs the UPF to cache the downlink data according to the AMF indication and sends the downlink data after the UE turns to the connected state.

2. EVENT NOTIFICATION SCHEME

In addition to supporting monitoring events such as UE Reachability and Availability Notification after DDN Failure already introduced by EPC, the "Downlink Data Delivery Status" event is also introduced.

(1) "UE Reachability" event reporting: Before sending data, the AF subscribes to the "UE Reachability" event through NEF to the network, and then sends downlink data when it receives the "UE Reachability" event reported by the network.

(2) "Availability Notification after DDN Failure" event reporting: the AF subscribes to the "Availability Notification after DDN Failure" event through NEF before sending the data. AF can send downlink data directly without waiting for network reporting. The AF can send downlink data directly without waiting for the network to report. If the UE cannot be paged, the AMF records that a DDN Failure has occurred, and then reports the UE availability to the AF after the UE has established a connection with the network.

(3) "Downlink Data Sending Status" event reporting: the AF subscribes to the "Downlink Data Sending Status" event through NEF before sending the data. The AF can send downlink data directly without waiting for network reporting. SMF reports to the AF according to downlink data sending result, including downlink data cache, downlink data drop, and downlink data sending success.

3.7.3.4 Coverage enhancement management

To effectively manage the use of the Coverage Enhancement mechanism by terminals, the 5G system can decide whether the UE is authorized to utilize Coverage Enhancement based on the contract information. The AF can query or modify the status of whether the UE uses coverage enhancement function via API. For eMTC access, the UE needs to report the coverage enhancement support capability (whether it supports CE mode B) to the AMF for subsequent control by the AMF based on the UE's coverage enhancement capability. For example, the UE configured as "voice-centric" is not allowed to enable CE mode B.

3.7.3.5 Small data overload control

1. SMALL DATA RATE CONTROL

The small data rate control feature allows the network to control the number of small data packets sent by IoT terminals, thus avoiding UPF/NEF overload, and includes the following rate control parameters:

(1) The number of packets allowed to be transmitted per unit of time.
(2) Whether exception reports are allowed to be sent after the rate control limit is reached.
(3) The number of exception reports allowed to be transmitted per unit of time.

2. SERVICE INTERVAL CONTROL

The service interval control feature is used for the network to control the minimum time interval for IoT devices to continuously send uplink small data packets. This feature is only applicable to delay-tolerant services. The service interval timer parameters can be saved in the contract information and sent to the UE by the AMF. During the operation time of service interval timer, the terminal is not allowed to send MO data or MO SMS but can send uplink exception reports or respond to network paging.

3. AMF/SMF OVERLOAD CONTROL

The AMF/SMF overload control function means that the AMF can request the RAN to reject the RRC connection request from the UE. The AMF/SMF can also initiate NAS-level congestion control.

4. Control plane (CP) data transmission evasion timer

The CP data transmission evasion timer means that when the network is overloaded, the AMF rejects the UE's CP data sending request and provides the UE with a CP data evasion timer. The UE cannot send small data through the control plane optimization scheme during the operation of the CP data evasion timer.

3.7.3.6 Reliable Data Service (RDS)

Reliable Data Service (RDS) is mainly for unstructured data transmission (non-IP data), and the RDS protocol is introduced to improve the reliability of unstructured data transmission.

RDS reuses the EPC mechanism to provide ACK messaging for non-IP data transmission between the UE-UPF or UE-NEF via the RDS protocol, as shown in figures 3-96 and 3-97 for the specific protocol stack.

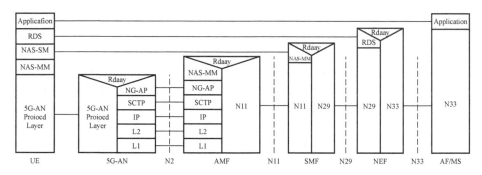

Figure 3-96: UE-NEF RDS protocol (for control plane data transmission)

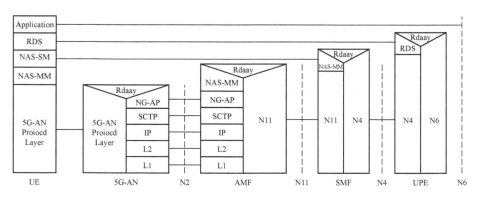

Figure 3-97: UE-UPF RDS protocol (for user plane data transmission)

3.7.3.7 Configuration of network parameters and UE behavior parameters based on the API interface

The AF is supported to provide network parameter configuration and UE behavior parameter configuration through a network capability open interface to optimize network performance.

The configuration of network parameters based on the API interface is performed as the AF sends the parameters of maximum latency, maximum response time, and suggested the number of downlink packets of the UE to the NEF, which in turn sends them to the corresponding AMF or SMF of the UE through the UDM. The Maximum Latency and Maximum Response Time can be used by the AMF to determine the power-saving function parameters of the UE, such as eDRX period; the Suggested Number of Downlink Packets is used to determine the number of cached packets in the extended caching scheme for high-latency communication.

Similar to the network configuration parameter process, the AF can also send the UE behavior parameters to the network through the API interface. See table 3-32 for the UE behavior parameters.

Table 3-32 UE behavior parameters

Expected UE behavior parameters	Description
Expected UE movement trajectory	Indicates the desired UE geolocation movement behavior, e.g., a pre-planned movement path. (optional parameter)
Fixed terminal indication	Indicates whether the terminal is a fixed terminal or a mobile terminal. (optional parameter)
Communication duration	Indicates how long the terminal remains in a connected state during data transmission, e.g., 5 minutes. (optional parameter)
Communication period	Indicates the time interval for periodic communication of the terminal, e.g., every one hour. (optional parameter)
Scheduled communication time	Indicates the time of the day or week when the UE can communicate with it, e.g., time: 13:00–20:00 every Monday. (optional parameter)
Battery indication	Indicates whether the terminal is sensitive to power consumption. (optional parameter)
Service characteristics	Indicates the type of data transmission by the terminal: single packet transmission (uplink or downlink), two packet transmission (one uplink packet and the corresponding downlink response packet), multiple packet transmission. (optional parameter)

Expected UE behavior parameters	Description
Scheduled communication type	Indicates the terminal's scheduled communication: uplink-only communication, downlink-only communication, or bi-directional communication. Can be used in combination with the scheduled communication time parameter. (optional parameter)

3.7.3.8 Event monitoring

The list of event monitoring functions is shown in table 3-33. They support the AF to obtain UE status information through the network capability open interface, such as UE reachability reporting and connection outage events, so that they can be processed according to the UE's status.

Table 3-33 Event monitoring functions

Event	Description	Event monitoring NF
Loss of connectivity	The network detects that it cannot communicate with the terminal	AMF
UE reachability	Indicates that the UE has changed to the connected state or page-monitoring state and can receive downlink data or SMS	AMF UDM:SMS Reachability
Location reporting	Indicates the terminal's current location or last known location	AMF, GMLC
Change of SUPI-PEI association	Indication of a change in the PEI associated with a Subscription Permanent Identifier (SUPI)	UDM
Roaming status	Indicates the current roaming status of the UE (current service PLMN and whether the PLMN is an HPLMN)	UDM
Communication failure	Determines communication failure with the terminal based on the RAN/NAS release reason value	AMF
Availability after DDN failure	Indicates that the UE becomes reachable when data transmission fails	AMF
Number of UEs present in a geographical area	Indicates the number of UEs in a geographic area described by the AF	AMF

Event	Description	Event monitoring NF
CN type change	Indicates the current CN type (EPC or 5GC) of a particular UE or group of UEs	UDM
Downlink data delivery status	Indicates the sending status of downlink data in the Core Network, downlink packet cache trigger event reporting, including downlink data cache event (also including estimated cache time), downlink data send success event, downlink data drop event	SMF

3.7.3.9 EPC-5GC interoperable unified northbound API interface

To enable the coexistence and evolution of NB-IoT/eMTC from EPC to 5GC, the 5G network is realized through the unified NEF/SCEF, and the external northbound API interface tries not to present the difference of the UE in EPC and 5GC.

For a UE that can access both EPC and 5GC, the network needs to select the SCEF+NEF co-deployment node for this UE. The SCEF+NEF co-deployment network architecture is shown in figure 3-98.

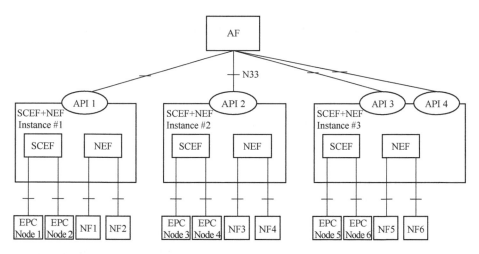

Figure 3-98 SCEF+NEF co-deployment network architecture

3.7.3.10 Idle-state inter-RAT mobility for NB-IoT

The 5G system supports idle-state inter-RAT mobility for NB-IoT access, i.e., idle-state reselection of terminals between NB-IoT access and eMTC, NR access. The

AMF determines the UE reselection from NB-IoT to other radio accesses based on the TAI where the UE is currently located (NB-IoT uses a dedicated TAI). Special processing on the network side includes the following:

(1) UE wireless capability processing: when the UE gets access through the NB-IoT, only NB-IoT wireless capability is reported to the network; the AMF also sends only UE wireless capability corresponding to the current access RAT of the UE to the NG-RAN.
(2) PDU session processing: the SMF decides whether to maintain, rebuild or release the PDU connection according to the contract information "PDU session continuity at inter-RAT mobility information," or base its processing on VPLMN policy.

3.7.3.11 NB-IoT QoS enhancement
NB-IoT QoS enhancement function means that NG-RAN can perform UE priority processing on control plane data transmission based on QoS signing information (NB-IoT UE Priority) provided by AMF, and the specific process can be found in the control plane optimization scheme uplink data transmission process.

3.7.3.12 Core Network node selection and redirection
The NG-RAN can access both EPC and 5GC at the same time and select the proper Core Network node based on UE access requirements. In addition, the network should be able to redirect the UE to other networks when the network is overloaded.

To support Core Network selection, the NG-RAN broadcasts the following information in system messages:

(1) whether connection to the 5GC is supported
(2) supported EPC IoT features
(3) supported 5G IoT features

The UE selects the CN type to be accessed according to the system message broadcast and the locally configured Preferred Network Behavior parameters. Then it indicates the CN type and the selected IoT features to the UE in the RRC access message. The RAN selects the corresponding CN node according to the UE instructions and completes the subsequent registration process.

The Core Network redirection scheme is used for the network to instruct the UE to redirect from the currently accessed network to another network using a NAS message-based indication scheme. From EPC to 5GC, the MME instructs the UE to redirect to 5GC in an Attach/TAU Reject message. From the 5GC to the

EPC, the AMF instructs the UE to redirect to the EPC in the Registration Reject message.

3.7.3.13 Unicast-based group messaging

The 5G network implements group messaging via unicast. The specific process is shown in figure 3-99.

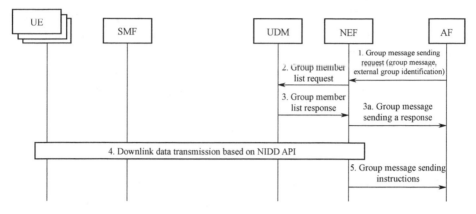

Figure 3-99 The process of the 5G network implements group messaging via unicast

Step 1: The AF sends a group message consisting of a request to the NEF, which includes the external group identifier and the pending group message.

Step 2: The NEF requests the corresponding group membership list from the UDM based on the external group identity.

Step 3: The UDM returns the list of group members to the NEF based on the pre-configured group information and the SMF information that serves the group members.

Step 3a: The NEF returns a response message to the AF indicating whether to accept the group message sending the request from this AF.

Step 4: The NEF sends the group message to the corresponding SMF according to the group member list in unicast mode, and the SMF sends it to the corresponding UE based on the control plane optimization scheme.

Step 5: The NEF sends a group message with instructions to the AF and returns the group message sending result.

3.7.4 Summary

The forward-looking design of the NB-IoT meets the main application requirements of 5G mass machine communication. 5G standardization smoothly connects NB-IoT terminals to the 5G Core Network and adapts and enhances them based on the 5G network architecture. NB-IoT is the air interface technology for 5G mMTC.

3.8 Telematics

3.8.1 Introduction

Cellular V2X (C-V2X) is a wireless communication technology for vehicles based on the evolution of cellular network communication technology, which supports network connectivity and applications between vehicles and vehicles, vehicles and roadside units (such as signals), vehicles and pedestrians, and vehicles and networks. Telematics is a key demand scenario for the application of 5G technology to vertical industries. C-V2X includes LTE-V2X and 5G-V2X.

In Rel-16, the C-V2X application scenarios consist of the following four main categories:

(1) Vehicle formation. Vehicle formation refers to the dynamic formation of a convoy of vehicles traveling in the same direction. The vehicles belonging to the convoy will obtain operational information from the head vehicle so that the distance between adjacent vehicles in the vehicle movement is closer, and thus reduce fuel consumption. This involves a series of complex application processes.

(2) Advanced driving. Advanced driving includes semi-automatic or fully automatic driving. Assuming long distances between vehicles, each vehicle or roadside unit sends its data sensed by sensors to the neighboring vehicles or roadside units, which allows the vehicles to operate together and plan a driving path. Vehicles also need to share their driving operations and destination information with neighboring vehicles. This ensures vehicle safety, avoids collisions, and improves driving efficiency.

(3) Remote driving. Remote driving refers to the remote operation of a vehicle by a driver or V2X application at a remote location (e.g., a passenger in the vehicle is unable to drive the vehicle for some reason, or the vehicle is located in a hazardous environment). Specifically, use cases can include vehicle driving scenarios where the route is more defined, such as a bus. Low-latency and high reliability are the most basic communication requirements for such scenarios.

(4) Sensor information sharing. Sensor information sharing allows vehicles, roadside units, application servers, or pedestrians to interact with data collected through local sensors. The vehicle can thus improve its perception of the surrounding environment and gain a panoramic view of the local

environment. In this scenario, high data transmission rates are a key communication requirement.

The 5G network telematics scheme includes network architecture enhancements for telematics services, Uu interface-based communication and QoS enhancements, PC5 interface-based communication and QoS enhancements, and interoperability with 4G systems to support the above-mentioned C-V2X application scenario.

The 5G-V2X scenario Uu interface and PC5 interface communication schematic is shown in figure 3-100. There are two main types of access technologies supported by C-V2X:

(1) Uu interface-based access technology. The data is sent through the Uu interface between the UE and the NG-RAN and relayed through the base station and Core Network to finally reach the destination vehicle or server.
(2) PC5 interface-based access technology. The data is sent through the PC5 interface between the UEs. The data will not pass through the Core Network, or even can be transmitted directly between vehicles without passing through the base station.

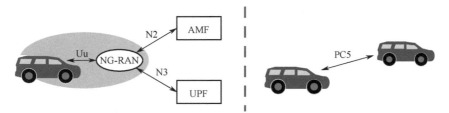

Figure 3-100 5G-V2X scenario Uu interface and PC5 interface communication

LTE-V2X research, mainly from the point of view of supporting vehicle safety, covers vehicle-to-vehicle and vehicle-to-network communication issues. Compared with LTE-V2X, 5G-V2X further researches the support of group communication including eV2X, QoS support enhancement, elements such as service authorization and parameter provision for PC5 interface, as shown in figure 3-101.

3.8.2 C-V2X architecture

In the C-V2X architecture, the network element or function that provides parameters to the V2X UE is called V2X Control Function, or V2X control function network element. In the LTE-V2X scheme, the parameters are provided by the V2X Control Function network element at the user plane. The architecture is shown in figure 3-102.

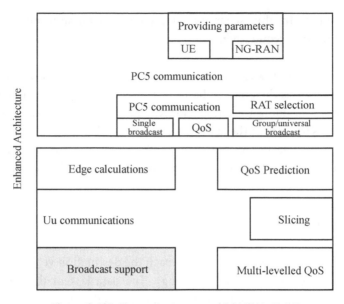

Figure 3-101 Research elements of 5G-V2X in Rel-16

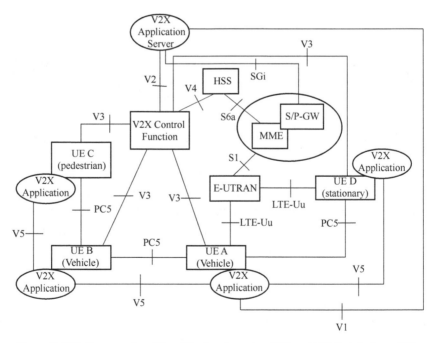

Figure 3-102 Non-roaming 4G architecture based on PC5 and LTE-Uu interfaces [16]

The architecture scheme of Rel-14/15 is the inheritance of the Proximity Service (ProSe) scheme. In the ProSe scheme, the network provides parameters to the UE through the ProSe functional network elements located at the user plane. For simplicity, this scheme is followed in the study of LTE-V2X, where the V2X control function network element is located at the user plane and provides parameters for V2X communication.

The C-V2X architecture scheme of Rel-16 is shown in figure 3-103.

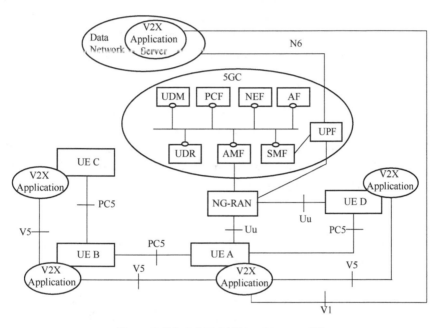

Figure 3-103 Rel-16 C-V2X architecture [17]

In the architectural scheme of Rel-16, the V2X control function network element becomes part of the PCF network element function to realize the function of the control plane network element to provide parameters. The main reasons are as follows:

(1) When deploying V2X control function network elements to user plane network elements, the processes of service authorization and access control are two separate processes, thus increasing the complexity of implementation. In addition, provisioning parameters need to be done through PDU sessions, which may incur additional latency and signaling overhead.

(2) PCF already supports providing policy information to the UEs, and V2X control function network elements deployed in PCF have fewer changes to the 5G standard.

3.8.3 Uu interface communication

Many applications, including remote driving, involve interactions between the UE and the V2X Application Server (see V2X Application Server in figure 3-103). These interactions are mainly implemented through Uu interface communication. In Rel-16, the Uu interface communication enhancement between the UE and the NG-RAN consists of two main aspects: first, the study of new QoS characteristics and the correspondence with the service demand characteristics; second, how the Core Network can effectively deal with this failure event when the QoS cannot meet the demand of V2X applications. In addition, based on the historical statistics of the Core Network, the predicted information of the QoS can be notified to the application server in advance for relevant processing.

3.8.3.1 Uu interface QoS features

The description of the telematics service requirements defined in 3GPP TS 22.186 [18] contains both security and non-security-related requirements for V2X. For example, the autonomous driving scenario has both safety-related V2X service requirements such as fleet driving and non-safety-related V2X service requirements such as map updating.

As mentioned earlier, the requirement for the telematics scenario is divided into four categories, including fleet, extended sensor information sharing, advanced driving, and remote driving. In addition, advanced driving contains six levels of autonomous driving, ranging from no autonomous driving at all to fully autonomous driving. The demand and autonomous driving levels will be described specifically in terms of dimensions including load (bytes), transmission rate (messages/second), maximum end-to-end delay (milliseconds), reliability (percentage), data rate (megabits per second), and minimum demand communication range (meters). Table 3-34 shows how these dimensions can correspond to the 5G QoS characteristics.

Table 3-34 V2X needs and the corresponding 5G QoS characteristics

V2X needs	Corresponding 5G QoS characteristics
Load	Can be associated with MDBV: MDBV can be defined as the sum of the maximum demand load and protocol overhead for Categories of Requirements and Level of Automation (LoA)
Maximum end-to-end delay	By definition, similar to PDB
Reliability	Results after combining PDB and PER
Data rate	Same as GBR QoS Flow or non-GBR QoS Flow

V2X needs	Corresponding 5G QoS characteristics
Minimum demand communication range	Special parameters of V2X that do not correspond to the QoS characteristics defined by 5QI

The QoS characteristics of some telematics scenarios can use the existing Delay Critical GBR 5QI values. For example, remote driving can correspond to a Delay Critical GBR service with a 5QI value of 85, and fleet can correspond to a Delay Critical GBR service with a 5QI value of 83, details described in 3GPP TS 23.501 [1] and 3GPP TS 22.186 [18].

3.8.3.2 Enhancing the Uu interface QoS notification mechanism

Many V2X applications are related to safe driving scenarios. If driving-related operations are not provided correctly, then the associated safety features cannot be met, which in turn can have serious consequences. For these reasons, V2X applications require faster and more accurate QoS Notification Control (QNC) mechanisms. When the QoS information that the network can provide changes, the network needs to be able to respond to this event faster. With these considerations in mind, Rel-16's V2X introduces multi-level QoS as a mechanism.

In the QoS notification mechanism of Rel-15 (see figure 3-104), since the Core Network element does not know the QoS parameters that the NG-RAN node can currently support when it receives a QoS notification control message, a more conservative control approach will be used to increase the probability that the NG-RAN nodes will meet the QoS requirements. However, this will result in a large gap between the adjusted policy and the current state of the NG-RAN, and it will take a longer time to match to the optimal policy.

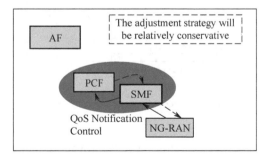

Figure 3-104 QoS notification mechanism for Rel-15

The core idea of the multi-level QoS mechanism (see figure 3-105) is to support the configuration of multiple levels of QoS requirements for an application on the NG-RAN side, and to create multiple QoS Profiles for these QoS requirements and bind them in the same QoS Flow. When the NG-RAN detects that the current QoS for this QoS Flow cannot be satisfied, it will use other QoS Profile bound in this QoS Flow.

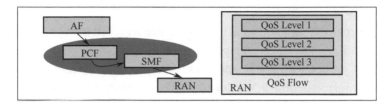

Figure 3-105 Diagram of the multi-level QoS mechanism

The flow of the multi-level QoS mechanism corresponding to figure 3-105 is shown in figure 3-106.

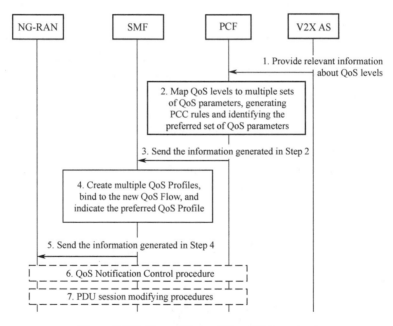

Figure 3-106 Flow of the multi-level QoS mechanism

Step 1: The V2X AS (i.e., AF) provides information about the QoS level to the PCF for the service, and the AF can map the contracted QoS values to the set of QoS parameters corresponding to each QoS level based on the Service Level Agreement (SLA). For multiple QoS levels, the AF identifies its preferred QoS level.

Step 2: The PCF maps the multiple QoS levels provided by the V2X AS to the corresponding sets of multiple QoS parameters, generates PCC rules, and identifies its preferred set of QoS parameters.

Step 3: The PCF sends the content generated in step 2 to the SMF.

Step 4: When the SMF receives the PCC rules corresponding to the multiple QoS parameter sets, it creates multiple QoS Profiles and binds these QoS Profiles to the newly created QoS Flow and identifies the priority of these QoS Profiles.

Step 5: The SMF sends this special set of QoS Profiles to the NG-RAN, which establishes the QoS Flow according to the preferred QoS Profile.

Step 6: Thereafter, if the QoS Profile of the current QoS Flow cannot be satisfied, then the NG-RAN notifies the Core Network using the QoS notification control mechanism. The NG-RAN can include the information of the currently supported QoS Profile in the notification message that informs the 5GC.

Step 7: Based on the information provided by the NG-RAN, the 5GC can initiate the corresponding PDU session modification process to adjust the QoS policy of the service flow.

3.8.3.3 V2X service enhancement based on QoS prediction

A V2X service may correspond to multiple configuration information, such as scenarios with different autonomous levels; each configuration will have different QoS requirements. When the QoS changes, V2X applications need to adjust their corresponding configurations accordingly.

The movement of the vehicle may cause the original QoS not to be guaranteed, as shown in figure 3-107. To enhance driving safety, some V2X applications need to be notified by the network in advance before QoS changes occur, e.g., remote driving, and autonomous driving scenarios.

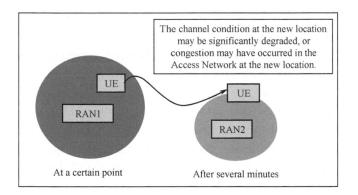

Figure 3-107 UE (vehicle) movement means that the original QoS cannot be guaranteed

V2X service enhancement based on QoS prediction is achieved by NWDAF network elements based on statistical information. The statistical information comes from the trajectory and QoS information provided by the application server, as well as the statistical information collected by the OAM approach. Specifically, the V2X application server sends a request to the 5GC containing the future location information of the UE, QoS information, and threshold information. Thereafter, based on the information provided by the application server and the OAM information, the NWDAF decides whether it is necessary to inform the V2X Application Server that the QoS may be changed. If necessary, the NWDAF network element informs the V2X Application Server about the relevant information about the QoS change through the NEF so that the V2X Application Server can adjust accordingly. The specific flow is shown in figure 3-108.

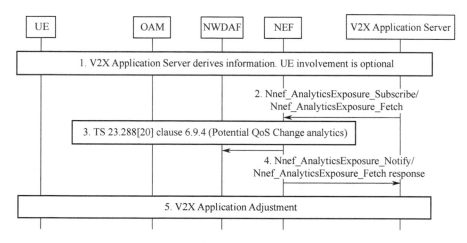

Figure 3-108 V2X service enhancement scheme based on QoS prediction [17]

Step 1: The V2X application server collects information about the UE, including the path of the UE, and QoS requirements.

Step 2: The V2X application server invokes the service message from the NEF, which includes the UE's future path information, QoS requirements, and threshold information.

Step 3: The NEF provides this information to the NWDAF, after which the NWDAF analyzes the information sent by the V2X application server.

Step 4: If the NWDAF decides that the V2X application server needs to be notified, then it notifies the analysis information about the QoS changes through the NEF's service.

Step 5: Thereafter, the V2X application server makes the appropriate adjustments. For example, the V2X application server can increase the distance between the V2X UEs or change the way of encoding, according to the deterioration of the QoS.

3.8.4 PC5 interface communication

The PC5 interface is the interface between the UE and the UE, making direct vehicle-to-vehicle communication possible. The PC5 interface mainly supports unicast, multicast, and broadcast communication, including support for the relevant signaling protocols, the definition of the relevant QoS models, and the establishment and maintenance of PC5 connections.

C-V2X PC5 communication includes LTE-PC5 communication and NR-PC5 communication. Rel-14/15 mainly provides broadcast communication features based on LTE-PC5, and Rel-16 provides unicast, multicast, and broadcast communication features for NR-PC5 interface communication. The unicast communication for PC5 interface involves the maintenance of connections, including operations such as connection creation, deletion, and update, as well as QoS negotiation based on NR-PC5.

The following contents describe the communication characteristics of the PC5 interface through three aspects: LTE-PC5 communication, NR-PC5 communication, and provisioning of PC5 parameters, respectively.

3.8.4.1 LTE-PC5 communication

The LTE-PC5 communication of V2X is enhanced on top of D2D for broadcast aspects, thus supporting the exchange of dynamic information (such as position, speed, and travel direction) between vehicles and the allocation of wireless resources.

The communication in the PC5 interface of LTE is based on a Per Packet ProSe Priority (PPPP), and a QoS model based on Per Packet ProSe Reliability (PPPR). The application in the UE carries the QoS requirements (PPPP and PPPR) of each packet in the packet and sends it from the application layer of the UE to the V2X layer. The V2X layer of the UE transforms the information from the application layer based on the configuration parameters or pre-configured information obtained from the V2X control network element, for example, mapping PPPP to delay information. Thereafter, the V2X layer sends the relevant information to the access layer of the UE for further execution.

The PC5 interface of LTE is enhanced only for the PC5 broadcast scenario. This mainly identifies and matches the identification information from the application layer for V2X services (i.e., V2X service ID) to the corresponding broadcast address (i.e., ID information of target layer 2). The identification and matching of parameters are mainly based on the mapping information issued by V2X control network elements or pre-configured mapping information.

For LTE-PC5 interface communication, the flow of its internal processing is shown in figure 3-109.

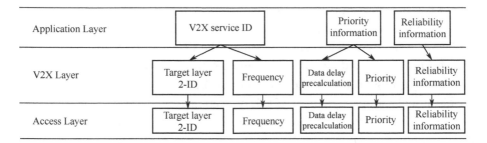

Figure 3-109 LTE-PC5 communication-related internal processing flow

3.8.4.2 NR-PC5 communication

Based on LTE-PC5, NR-PC5 adds unicast and multicast communication scenarios and enhances the QoS model for unicast, multicast, and broadcast according to the needs of the new scenarios.

PC5's QoS management reuses Uu's QoS Flow-based model. For multicast and broadcast, the same QoS Flow-based model is used in order to simplify the UE implementation. In addition, PC5 unicast communication requires the establishment of PC5 connections between the UEs. The maintenance of the connections involves the creation, update, and deletion of PC5 connections.

PC5 QoS Flow is the finest-grained QoS differentiation of the same PC5 connection (i.e., a connection represented by source-destination Layer 2 ID information). When the V2X layer receives service data or transport requests from the application layer, it first determines whether there is an existing PC5 QoS Flow that matches the service data or request (i.e., QoS rules based on the existing PC5 QoS Flow). If there is no PC5 QoS Flow matching the service data or request, the V2X layer generates PC5 QoS parameters based on the V2X application requirements or V2X service type (e.g., PSID or ITS-AID) provided by the V2X application layer, either by the PCF or by locally configured mapping relationships. The UE creates a new PC5 QoS Flow for the generated PC5 QoS parameters. Thereafter the UE assigns a PFI to this PC5 QoS Flow and generates a PC5 QoS rule for this PC5 QoS Flow. The V2X layer maintains a context for each PC5 connection, which includes information such as the mapping of PFI to PC5 QoS parameters and V2X service type. Figure 3-110 depicts the processing flow within the PC5 communication UE; the right-hand side of the figure depicts the reuse of existing QoS Flows, which may update the corresponding PC5 connection context.

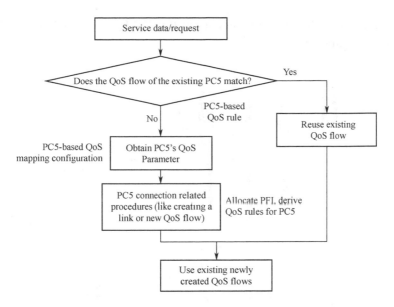

Figure 3-110 Processing flow inside PC5 communication UE

When the UE assigns a new PFI to a V2X service, the UE saves the corresponding QoS context and the corresponding QoS rule. When the UE releases the PFI, the UE removes the corresponding PC5 QoS context and PC5 QoS rule. The PC5 QoS rule contains the PFI of the associated PC5 QoS Flow, the priority information, and the PC5 Packet Filter Set (PFS). The V2X layer ensures that V2X services associated with different frequencies (depending on the definition of ITS) are assigned to different PC5 QoS Flows. The QoS-related information associated with the PC5 connection is shown in figure 3-111. During the PC5 connection establishment process, the V2X layer provides the PFI, PC5 QoS context, and Layer 2 ID information to the AS layer.

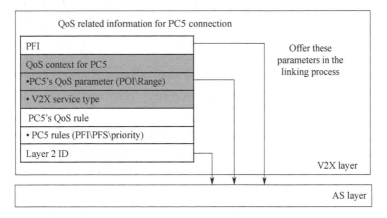

Figure 3-111 QoS-related information for PC5 connections

A summary of QoS Flow processing for the PC5 QoS rule in 3GPP TS 23.287 is shown in figure 3-112.

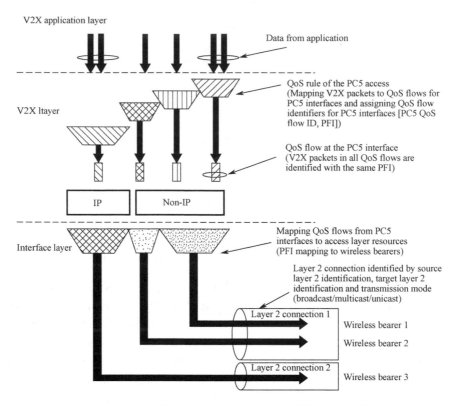

Figure 3-112 QoS Flow processing based on PC5 QoS rule [17]

For PC5 unicast communication, PC5 connections are required between the UEs. Maintenance of PC5 connections involves operations such as the creation, update, and deletion of PC5 connections. The guidelines for PC5 connection management are as follows:

(1) PC5 connections support one or more V2X services in a pair of UEs. All V2X services transmitted using the same PC5 connection use the same application layer ID.
(2) For data from a V2X application, if its application layer ID and network layer protocol are the same as the application layer ID and network layer protocol of an existing PC5 connection, this PC5 connection is reused to transmit data from this V2X application.
(3) After successful PC5 unicast link building, the UE uses the same pair of layer 2 IDs for subsequent PC5-S signaling interactions and V2X service

data transmission. The V2X layer of the sending UE indicates to the AS layer whether it is a PC5-S signaling message or a V2X service data transmission.

(4) For each PC5 connection, a unique PC5 connection identifier is assigned within the UE to uniquely identify the PC5 connection in the UE. Each PC5 connection is associated with a corresponding Unicast Link Profile. The contents the Unicast Link Profile contains are shown in figure 3-113.

Unicast Link Profile
Service type
Application layer ID
Layer 2 ID
Network layer protocol
Set of QoS flow identifiers and QoS parameters for PC5 of each V2X service

Figure 3-113 PC5 unicast connection description information

An example of PC5 unicast connections is shown in figure 3-114, depicting the correspondence between the application layer identification and PC5 connections. In the figure, the UE A and UE B have two PC5 connections corresponding to application layer ID 1 of UE-A and application layer ID 2 of UE-B, application layer ID 3 of UE-A, and application layer ID 4 of UE-B, respectively.

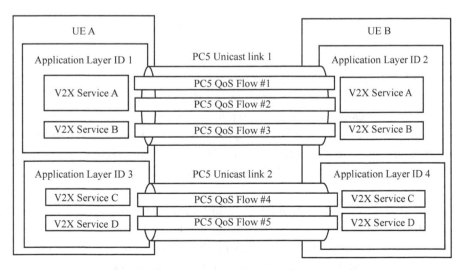

Figure 3-114 Example of a PC5 unicast connection [17]

The PC5 connection establishment process is shown in figure 3-115. There are two ways to establish PC5 connections for unicast, one is the UE-oriented PC5 connection establishment, and the other is the V2X service-oriented PC5 connection establishment. The UE-oriented PC5 connection establishment is mainly a scenario where the identification information of the receiver is known at the application layer, while the V2X service-oriented PC5 connection establishment is a scenario where the UE is interested in a specific V2X service is searched for by means of a broadcast.

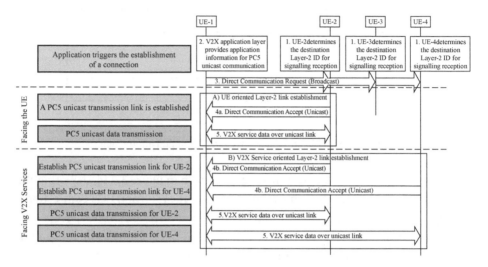

Figure 3-115 PC5 connection establishment process

The flow of the establishment of a UE-oriented layer 2 connection is outlined as follows:

(1) The UE-1 sends a Direct Communication Request message via a broadcast message with a broadcast address layer 2 ID associated with the V2X service type. The message contains the application layer identity of the target UE (e.g., UE-2), which is used by the upper layer of the target UE to identify this message and send a response message. The source layer 2 ID of the message is the source layer 2 ID of the UE-1, and the source layer 2 ID is assigned by the UE-1.

(2) The Direct Communication Request message uses the default AS layer setup information to enable the relevant target UE to parse this message.

(3) The UE-2 replies to this message using the source layer 2 ID of the message as the destination layer 2 ID and using its own source layer 2 ID as the source layer 2 ID. The UE-1 obtains the layer 2 ID of the UE-2 for future communication.

The User Info in the Direct Communication Request message contains the upper layer application identification information of the source UE in addition to the upper layer application identification information of the target UE.

The flow of the UE-oriented Layer 2 connection establishment is as outlined follows:

(1) The UE-1 contains in the Direct Communication Request message information for the layer 2 connection establishment (e.g., service informative information) for the other UEs to determine whether to respond to this request.

(2) The UEs interested in this V2X service respond to this message (e.g., UE-2 and UE-4).

The PC5 connection update flow is shown in figure 3-116. Updates are mainly used to add a new V2X service to an existing PC5 connection, to remove a V2X service from an existing PC5 connection, or to update the PC5 QoS Flow in an existing PC5 connection.

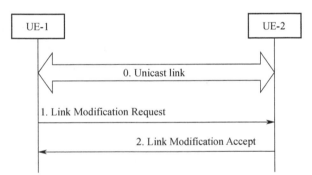

Figure 3-116 PC5 connection update process

Step 1: The V2X application layer in the UE-1 provides the application layer information for PC5 unicast communication, which includes the service type of the V2X application and the application layer identification information of the originating UE. If UE-1 decides to reuse the existing PC5 connection, then UE-1 sends a link modification request to the UE-2, which may contain QoS information, V2X service information.

Step 2: The UE-2 replies with a link modification acceptance message, which may contain QoS information, V2X service information Thereafter, the V2X layer passes the updated information to the AS layer for AS to modify the connection-related context.

The PC5 connection deletion process is shown in figure 3-117 and is mainly used to release PC5 connections.

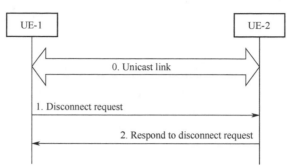

Figure 3-117 PC5 connection deletion process

Step 1: The UE-1 sends a disconnection request message to the UE-2 to release the PC5 connection and delete all context data associated with the PC5 connection.

Step 2: After receiving the disconnection message, the UE-2 responds to the disconnection request. On the other hand, the UE-2 locally deletes all context data associated with the PC5 connection. The V2X layer of the UE-1 and UE-2 notifies the AS layer that the PC5 connection has been deleted, which enables the AS layer to delete the associated context.

3.8.4.3 Provision of PC5 parameters

The parameters for PC5 communication can be pre-configured in the Universal Integrated Circuit Card (UICC) or Mobile Equipment (ME) or provided over the network. For system interoperability, the PCF of the 5G system is not required to interact with the V2XCF of the EPC when the UE moves between the EPC and the 5GC, considering that the V2X parameters are session independent. In order to minimize the impact on the EPC, it is assumed in Rel-16 that parameters involving 5GS communication will not be provided through the V2XCF.

1. PROVISION OF INITIAL PARAMETERS

At initial registration, the UE that supports V2X provides V2X capability information for the AMF to find out the relevant PCF via NRF or local configuration. If the UE needs to obtain V2X parameters, then the UE needs to include the indication information for providing V2X parameters in the UE Policy Container. After the AMF selects the PCF, the UE's capability information and contract information is provided to the PCF along with the UE Policy Container. Based on this information, the PCF determines the need to provide information about the parameters associated with PC5. The latest parameters are then provided to the UE through the UE Configuration Update process.

2. UPDATE OF PARAMETERS

The UE initiates a parameter update process when the timer associated with the UE-related parameters expires, or when there is no locally stored parameter available in the current area, or when other unusual problem situations occur. This process is implemented through a UE configuration update process triggered by the UE. In addition, when the UE moves (e.g., from one PLMN to another), the PCF may also update the relevant parameter information of the UE. The PCF may also need to update the relevant parameters sent to the UE when the contract information of the UE is changed.

3.8.5 Summary

C-V2X has received increasing attention in recent years thanks to the deep integration of the automotive, transportation, and communication industries. C-V2X enhances the communication of the Uu interface, PC5 interface, and network intelligence by adapting Rel-16, which has significance for improving communication efficiency, and thus can alleviate the current pressure of safety, congestion, and pollution faced by the transportation field.

3.9 Ultra-Reliability and Low Latency Communication

3.9.1 Introduction

Ultra Reliable and Low Latency Communication (URLLC) is the main enabling technology of 5G for vertical industries and some demanding service scenarios. Typical URLLC service scenarios are described below:

(1) Smart manufacturing. Smart manufacturing is oriented to the full lifecycle of products. Through smart perception, human-computer interaction, decision-making and execution technology, a smart design process, manufacturing process, and manufacturing equipment can be realized, including motion control, controller to controller communication (C2C communication), mobile robots, and industrial AR. These industrial control scenarios are extremely sensitive to indicators such as the latency and reliability of communication. In some demanding scenarios, end-to-end latency is required to be less than 1ms, and its reliability should be as high as six 9s to eight 9s (i.e., the packet loss rate is less than 10^{-6} to 10^{-8}).

(2)   Autonomous vehicle. For example, remote driving/control of forklifts/cranes and other construction machinery, ferries at airports/ports/docks/parks, autonomous trucks on expressways, and autonomous taxis in formation. These high-speed vehicles have high safety requirements. Higher speed means a lower latency requirement for the communication system in order to complete instruction transmission and execution within a certain moving distance. Therefore, autonomous vehicle is a typical scenario with high reliability and low latency.

(3)   Smart grid. Smart power grid applications (e.g., distributed feeder automation and precise load control) require a very high end-to-end response time of the system, and thus a stricter demand for latency and reliability of instruction transmission. Typical end-to-end transmission latency is required to reach around 10ms, and reliability requires five 9s to six 9s.

(4)   Cloud AR/VR and real-time games. These services require large bandwidth and low latency to improve user experience.

The common requirements of the above applications on the transmission network are mainly latency and reliability. Generally speaking, the URLLC service requires a delay of 10ms or less, and a packet loss rate of 10^{-5} for reliability.

From the point of view of system design, low latency and high reliability are two mutually affecting indexes. If we only consider one index, we can apply simpler techniques to achieve it. For example, if we only need high reliability, we can apply retransmission mechanisms such as the application layer. If low latency is the only consideration, we can achieve it by doing whatever we can in transmission. However, many high-value services often need to meet the requirements of both low latency and high reliability at the same time, which is a hugely demanding challenge for system design.

3.9.2 Ultra-reliability transmission

Network reliability refers to the reliability of network devices and paths. For around 99.999% or 99.9999% of the reliability requirements, single-link transmission requires high device reliability of a single processing node on the transmission path. In addition, if the transmission network does not support transmission technologies such as the Time Sensitive Network (TSN) or Deterministic Network (DetNet), due to the occurrence of parallel flow in the case of single link, short-term micro-granularity congestion may occur on the switch of the transmission network, resulting in packet transmission exceeding the delay threshold, which reduces service availability. Based on the reliability performance of existing devices and links, single-link transmission is difficult to meet the reliability of five 9s or six 9s. A common practice in the industry is to adopt dual transmission to reduce the high demand for single-point reliability. For example, IEEE TSN networks adopt Frame Replication and Elimination for Reliability, as defined by 802.1CB [19], which function to achieve multi-channel redundant transmission, to ensure high reliability requirements of services.

Similarly, in the 5G network, in order to avoid reliability problems and occasional micro-congestion caused by a single link, three different schemes for ultra-reliability transmission through redundant link transmission have been designed.

3.9.2.1 High reliability transmission based on redundant sessions

Figure 3-118 shows the redundant session transmission scheme. Based on Dual Connectivity (DC) redundant session transmission, it adopts the mode of redundancy packet transmission by Dual PDU session to realize high reliability transmission. The UE connects to two RAN nodes by DC and establishes two PDU sessions with two anchor points on user plane, namely UPF 1 and UPF 2, through different RAN nodes. Based on these two PDU sessions, the 5G network provides the UE with two independent redundant paths for transmitting the same packet.

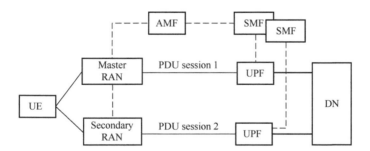

Figure 3-118 Redundant session transmission scheme

The key to implementing the redundant session transmission scheme is to ensure that two sessions can select different RAN nodes and UPF nodes respectively. In addition to the reliability of the control plane, different SMFs need to be selected for the two sessions. These choices can be made through the different DNN and S-NSSAI combinations as well as network configurations provided by the UE during session configuration.

Specifically, two different URSP rules can be placed on the UE to map redundant packets sent by the application layer to different DNN and S-NSSAI combinations. Thus, the UE carries different DNN and S-NSSAI when initiating session establishment. Then, the AMF determines whether to select different SMFS for the two sessions based on the DNN and S-NSSAI and local configurations. Similarly, SMF selects different UPFs based on local configuration.

Based on the S-NSSAI and DNN associated with PDU sessions, user contract information, and local policies, SMF determines the redundancy transfer requirements of PDU sessions on the user side to the NG-RAN through the RSN (Redundancy Sequence Number) parameter of the session granularity. According to RSN parameters indicated by SMF and local configuration, the NG-RAN node determines the redundant transmission resources that should be provided to user plane. The NG-RAN establishes DC connection for the UE as required and selects RAN side nodes for two sessions according to the RSN, and distributes the two sessions on different RAN nodes. In this way, the transmission between RAN and UPF can be independent of each other.

3.9.2.2 High reliability transmission based on redundant service flows
Figure 3-119 shows the redundancy service flow transmission. For some scenarios, the bottleneck of end-to-end reliability may exist in the backhaul network based on the redundant transmission of the Packet Data Convergence Protocol (PDCP) layer and the resilience of the control plane. Due to the distance of the backhaul network and the complexity of the deployment, the reliability of a single N3 tunnel may not meet the requirements of URLLC services. In this case, redundant transport can be deployed between the anchor UPF and the NG-RAN nodes to improve the reliability of the backhaul network by establishing two independent N3 tunnels for a single PDU session.

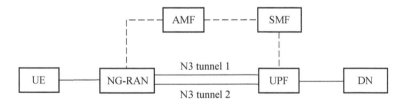

Figure 3-119 Redundancy service flow transmission

To ensure that the two N3 tunnels are transmitted through separate transport layer paths, different routing information (such as different IP addresses or network instances) must be provided in the tunnel information. Then, based on the network deployment and configuration, the routing information can be mapped to different transport layer paths.

When establishing a QoS flow for URLLC, if the SMF decides to perform N3 redundant transmission of this QoS Flow based on the authorized 5QI and the NG-RAN node's capability and carrier configuration, the SMF will notify the anchor UPF and the NG-RAN node through an N4 interface and N2 information respectively. Two pieces of independent tunnel information will be allocated simultaneously on the NG-RAN and the UPF to create redundant N3 transport tunnels.

For each downlink packet of this QoS flow received from the DN by the anchor UPF, the anchor UPF copies the packet and allocates the same GTP-U sequence number for redundant transmission. These packets are distributed to the NG-RAN nodes over two redundant N3 tunnels. The NG-RAN node then resends the packets according to the GTP-U sequence number, removes the duplicates, and forward them to the UE. For each uplink packet of this QoS flow received by the NG-RAN from the UE, the NG-RAN node copies the packet and allocates the same GTP-U sequence number, and the anchor UPF forward the packet to DN after the de-duplication process according to the GTP-U sequence number.

Two I-UPFs may be inserted between the NG-RAN and the anchor UPF to support redundant transmission of the N3 and N9 at the same time. In this scenario, the I-UPF only forward uplink and downlink packets; replication and de-duplication of packets are still performed by the NG-RAN and the anchor UPF.

3.9.2.3 High reliability transmission based on transport layer redundancy

Redundant transport on the transport layer does not require the establishment of a redundant GTP-U tunnel on the N3 interface but is implemented by providing two redundant transport layer paths between the UPF and the NG-RAN. During session establishment, the SMF selects a UPF that supports redundant transmission at the transport layer as the session anchor UPF and establishes an N3 GTP-U tunnel between the UPF and the NG-RAN. For downlink data transmission, the UPF sends downlink data packets over the N3 GTP-U tunnel and copies downlink data at the transport layer, while NF-RAN is in charge of de-duplicating the downlink data packets at the transport layer and sending them to the UE. For uplink data transmission, the NG-RAN sends uplink data packets over the N3 GTP-U tunnel and replicates the uplink data at the transport layer, and then UPF de-duplicates the uplink data packets at the transport layer and forward them to the DN.

The transport-layer redundant transmission mechanism is not standardized in 3GPP. It is up to the network deployment to decide which standardized or non-standard technologies to implement.

3.9.3 PDB decomposition

Packet Delay Budget (PDB) defines the upper limit of Packet Delay between the UE and the anchor UPF. The end-to-end PDB is composed of the CN PDB of the Core Network and the AN PDB of the air interface. CN PDB refers to the delay budget between the RAN node and the anchor UPF. AN PDB refers to the delay budget of air interface transmission.

In actual network deployment, carriers may choose different UPF deployment locations based on different service requirements and deployment environments. For example, the UPF can be deployed in a high position to achieve a larger coverage. The UPF can also be deployed near the RAN to save CN PDB and increase AN PDB for flexible wireless resource scheduling on the RAN side. In pre-5G networks, the decomposition of the delay budget between the Core Network and air interface is relatively fixed. However, for low-latency services, due to the strict end-to-end delay budget, it is necessary to carry out fine control in the network. Figure 3-120 shows the end-to-end PDB decomposition.

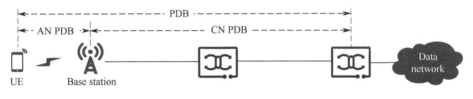

Figure 3-120 End-to-end PDB decomposition

5G supports the dynamic PDB decomposition scheme based on the deployment location of NEs. That is, for low-latency GBR 5QI, CNPDB can be configured on the SMF or NG-RAN nodes based on different anchor UPFs and NG-RAN nodes. If it is configured on SMF, the SMF sends the dynamic CN PDB value to the NG-RAN node when the QoS flow is established, or the dynamic CN PDB value of the QoS flow changes (when activities such as switching or inserting I-UPF occur). Based on the received CN PDB, the NG-RAN can more accurately estimate the available AN PDB for the air interface to achieve more accurate scheduling.

3.9.4 QoS monitoring

URLLC services have high QoS requirements. Therefore, to ensure delay and reliability in applications, the network must be able to sense link quality in time and adjust it accordingly. By measuring the packet delay in real time, the QoS management of URLLC services can be visualized, and real-time SLA detection of slicing, QoS closed-loop warning of ultra-reliability services, and traffic realization ability can be improved.

Core Network QoS monitoring can be realized by flow granularity monitoring or node granularity monitoring.

3.9.4.1 Real-time QoS monitoring for QoS flow

The delay monitoring of QoS flow granularity is based on actual service packets between the UPF and UE. Uplink/downlink packet delay consists of uplink/

downlink packet delay between the UE and the NG-RAN node, and uplink/ downlink packet delay between the NG-RAN node and the anchor UPF.

The SMF sends QoS monitoring policies for QoS flows to the UPF through the PDU session establishment or modification process. Based on the received QoS regulation policy, the anchor UPF initiates a packet delay measurement between the NG-RAN node and the anchor UPF, and the NG-RAN node-initiates a packet delay measurement of the uplink/downlink packet delay of the Uu interface. If all user plane nodes of the 5G system are time synchronized when the anchor UPF sends downlink packets, the anchor UPF adds a timestamp to downlink packets. The NG-RAN calculates the downlink delay based on the time of receiving the packet and sends the downlink delay and Uu interface delay to the UPF anchor in the uplink packet. In the absence of an uplink service packet, the NG-RAN node can send a dummy uplink packet to the UPF as a monitoring response packet. The anchor UPF can calculate each transmission delay of uplink and downlink packets based on the information reported by the NG-RAN and the sending time of uplink packets.

If the UPF and the NG-RAN do not support time synchronization, the anchor UPF will record the local time when the downlink packets are sent. Then NG-RAN node will provide the UPF with the Uu interface's uplink/downlink packet delay measurement results and local packet receiving and sending times through the N3 interface. The anchor UPF can then calculate the packet loopback time based on the local time of receiving uplink packets and the information reported by NG-RAN.

The anchor UPF can report QoS monitoring results to the SMF based on certain conditions (e.g., reaching the SMF reporting threshold) for application layer alarms or other QoS policy decisions.

3.9.4.2 Node-level QoS monitoring

Node-level QoS monitoring can provide delay estimation results of node granularity. Different from the QoS monitoring at QoS flow level mentioned above, node-level monitoring is based on the GTP-U Echo request/response in the user plane transmission path to estimate the delay of Core Network packets. For details, please refer to 3GPP TS 28.552 [14].

3.9.5 Summary

5G network support for URLLC features includes an ultra-reliability transmission mechanism based on a redundant session, tunnel or transport layer, dynamic decomposition of PDB, and QoS monitoring. Based on the above characteristics, the 5G Core Network can provide ultra-reliability transmission beyond the performance of a single link and can carry out more accurate control and real-time measurement of transmission delay, so as to carry services with demanding QoS performance requirements.

3.10 Convergence of fixed and mobile networks

3.10.1 Introduction

ITU-T defines the convergence of fixed and mobile networks as the ability to provide services or applications to terminal users in a given network, regardless of fixed/mobile access technology and user location. It is often referred to as fixed-mobile convergence.

Fixed-mobile convergence is an important feature of 5G networks. The 3GPP SA1 working group listed "supporting fixed network services through mobile services" as one of the requirements of 5G in the early stage of 5G standardization in 2015. In Rel-15, 3GPP has completed the development of standards for terminal access to the 5G networks through untrusted non-3GPP access technologies. In Rel-16, 3GPP and BBF jointly completed the standard formulation of terminal access to the 5G network through wireline access technology, and 3GPP also completed the standard formulation of terminal access to the 5G network through trusted non-3GPP access technology.

5G fixed-mobile convergence includes four parts: fixed wireless access, non-3GPP access, hybrid access, and IPTV service support:

(1) "Fixed wireless access" means the provision of network connectivity to fixed network terminals through a mobile network. In areas where fixed-line deployment is difficult, or bandwidth is insufficient (such as some rural areas where fiber is geographically or financially difficult to deploy), the technology could provide high-bandwidth 5G connections to fixed terminals.

(2) "Non-3GPP access" comprises three types of non-3GPP access technologies: untrusted non-3GPP access, trusted non-3GPP access, and wireline access. Terminals can access 5G networks through these three non-3GPP access technologies. Under the design principle of "access independent," the 5G network controls 3GPP access and non-3GPP access simultaneously by using a unified Core Network and protocol mechanism, reducing network complexity and cost related to operation and maintenance.

(3) "Hybrid access" refers to the state in which terminals access the 5G network through both 3GPP access technology and non-3GPP access technology. It significantly improves network bandwidth, and also ensures the continuity of terminal services when one of the access technologies is unavailable. Hybrid

access is of great significance for 5G networks to support large-bandwidth services (such as AR/VR services).

(4) "IPTV service support" means that terminals can access and obtain IPTV service functions through 5G networks. It is also a necessary enabling technology for 5G networks in home application scenarios. The support of 5G networks for IPTV services is of great significance, for it means that 5G fixed-mobile convergence has realized the convergence of network access, and also the convergence of network services.

3.10.2 Overall architecture of 5G fixed-mobile convergence

Figure 3-121 shows the 5G fixed-mobile convergence architecture. In 5G fixed-mobile convergence architecture, terminals can access 5G networks through both 3GPP access technology and non-3GPP access technology:

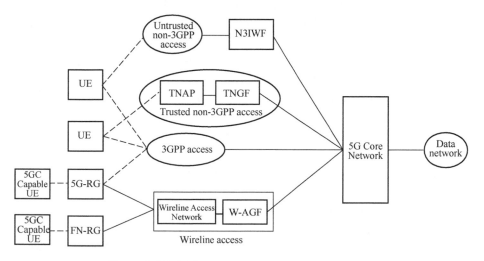

Figure 3-121 5G fixed-mobile convergence architecture

(1) Fixed wireless access. The UE of mobile networks and the 5G Residential Gateway (5G-RG) of fixed network terminals that support 5G protocol can both access 5G networks through 3GPP access technology. These types of access are called Fixed Wireless Access. See section 3.10.3.1 of this book for details.

(2) Non-3GPP access. Both mobile network terminals and fixed network terminals can access 5G networks through non-3GPP access technology, as follows:

a. A mobile network terminal UE can access the 5G network through untrusted non-3GPP access technology. See section 3.10.3.2 of this book.

    b. A mobile network terminal UE can access the 5G network through trusted non-3GPP access technology. See section 3.10.3.3 of this book.

    c. Fixed Network terminals such as 5G-RG, Fixed Network Residential Gateway (FN-RG), and 5G Capable UE can access the 5G network through wireline access technology. See section 3.10.3.4 of this book for details.

(3) Hybrid access. The state in which terminals (including UE and 5G-RG) are connected to the 5G network through both 3GPP access technology and non-3GPP access technology is called hybrid access. See section 3.10.3.5.

Under the "access independent" design principle, 5G networks establish a unified management and control framework for 3GPP and non-3GPP access, including basic features such as mobility management, session management, QoS mechanism, and policy control. See section 3.2 of this book.

It is important to note that the connection status of UE/RG under 3GPP and non-3GPP access is different. The UE/RG in 3GPP access mode changes from the connected state to the idle state mainly to save the context storage resources and terminal power of the Access Network, while the UE/RG in non-3GPP access mode changes from a connected state to an idle state mainly because the physical connection of the Access Network is disconnected. That is, the UE/RG will always be connected if the physical connection of the non-3GPP Access Network can be maintained. Of course, if the physical non-3GPP connection is restored within a certain period, the UE/RG will be restored. RG here includes 5G-RG and FN-RG.

3.10.3 Key features of 5G fixed-mobile convergence
3.10.3.1 Fixed wireless access
5G-RG connects to the fixed wireless access architecture of the 5G network through 3GPP access technology, as shown in figure 3-122.

The basic features of fixed wireless access are the same as those of a UE accessing the 5G network through 3GPP access technology. See the basic features of 5G networks in section 3.2 of this book.

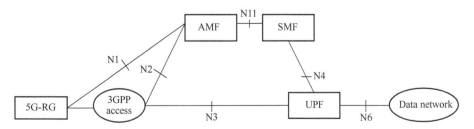

Figure 3-122 Fixed wireless access architecture

3.10.3.2 Untrusted non-3GPP access

3.10.3.2.1 Untrusted non-3GPP access architecture

The UE uses untrusted non-3GPP access technologies to access the 5G network, as shown in figure 3-123.

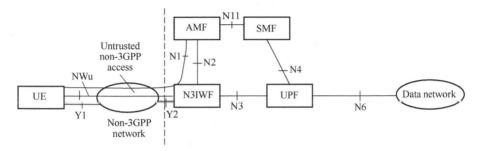

Figure 3-123 Untrusted non-3GPP technology's access to the 5G network

Untrusted non-3GPP access refers to non-3GPP access technologies that are not trusted by operators, such as Wi-Fi deployed by users. It is the signal access point of the UE and the first hop node for the UE to access the 5G network.

N3IWF is an untrusted non-3GPP access gateway deployed by the operator. Its network topology position is equivalent to that of the NG-RAN when the UE accesses the 5G network through 3GPP access technology. It supports N2 and N3 interfaces in the 5G Core Network, and it can transfer NAS signaling between the UE and AMF.

Figures 3-124 and 3-125 show the control plane protocol stacks for untrusted non-3GPP access before and after the establishment of the signaling Internet Protocol Security (IPSec) tunnel.

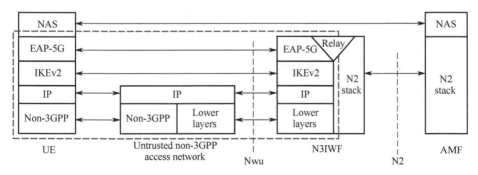

Figure 3-124 Control plane protocol stack of untrusted non-3GPP access before a signaling IPSec tunnel is established

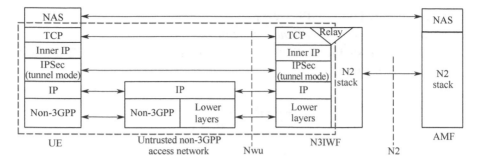

Figure 3-125 Control plane protocol stack for untrusted non-3GPP access after a signaling IPSec tunnel is established

Compared with 3GPP access control plane protocol stack, the Core Network control plane protocol stack and NAS protocol layer are the same as those of 3GPP access control plane protocol stack and NAS protocol layer. The difference mainly lies in the protocol stack on the Access Network side, as shown in the dotted boxes in figures 3-124 and 3-125.

(1) In figure 3-124, in the registration process and the service request process initiated by the UE after it enters the idle state, the UE and N3IWF will establish an Extensible Authentication Protocol (EAP-5G) session based on the protocol layer of Internet Key Exchange Version 2 (IKEV2). The session is used in both processes to transmit the AS parameters and NAS signaling that the UE has been accessed by the 5G network through non-3GPP access technologies. EAP-5G is a signaling session between the UE and N3IWF and is introduced to support the UE to access the 5G network through untrusted non-3GPP access technology. It is subsequently also used for trusted non-3GPP access and wireline access.

During the execution of these two processes, the UE and N3IWF establish an IPSec tunnel for the transmission of NAS signaling on the Access Network side, which is used to carry NAS signaling in other processes. Because the IPSec tunnel is specially used to carry NAS signaling, it is called the signaling IPSec tunnel.

(2) As shown in figure 3-125, except for the registration process and the service request process initiated by the UE after it enters the idle state (such as the degradation process and session management process), NAS signaling is carried by the signaling IPSec tunnel for transmission on the Access Network side.

There is also a layer of Transmission Control Protocol (TCP) and a layer of Inner IP Protocol between the UE and N3IWF. The TCP Protocol layer is used to ensure the reliable transmission of NAS signaling between the UE and N3IWF, while the Inner IP protocol layer serves as the Inner IP of the "IP in IP" transmission mechanism of IPSec in tunnel mode.

There was a lot of discussion in Rel-15 about whether 5G will introduce EAP-5G sessions. There were two main views. One was to introduce EAP-5G to transmit AS parameters and NAS signaling on the non-3GPP Access Network side. The main rationale behind this was that EAP-5G can be used as a general mechanism for all non-3GPP, instead of untrusted non-3GPP only, to transmit AS parameters and NAS signaling on the Access Network side. It would ensure good compatibility and scalability. Another was not to introduce EAP-5G, while directly extending and transmitting AS parameters and NAS signaling on IKEV2 messages, for the introduction of EAP-5G scheme may affect EAP protocol, which needs to be re-standardized to IETF, and thus may delay the progress of 3GPP Rel-15. Eventually, 3GPP accepted EAP-5G due to its technical advantages. As was the driving force behind the original introduction of EAP-5G, other non-3GPP access technologies do use EAP-5G to transmit AS parameters and NAS signaling, as described in sections 3.10.3.3 and 3.10.3.4 of this book.

Figure 3-126 shows the user plane protocol stack of untrusted non-3GPP access.

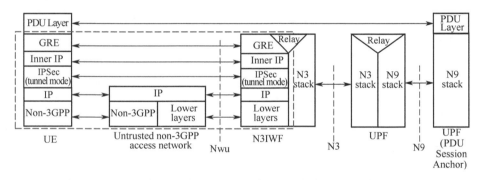

Figure 3-126 User plane protocol stack of untrusted non-3GPP access

Compared with the user plane stack that 3GPP accesses, the user plane stack of the Core Network is the same as that of 3GPP. The main difference lies in the protocol stack on the access side (as shown in the dotted box in figure 3-126):

(1) Data on the user side is encrypted through IPSec tunnel protocol layer on the untrusted non-3GPP Access Network side and is transmitted through DRB on the 3GPP Access Network side.

(2) There is also a Generic Routing Encapsulation (GRE) protocol and a layer of Inner IP protocol between the UE and N3IWF. GRE protocol is used to

carry QFI and mask the differences of application layer protocols to support Ethernet PDU session types and unstructured PDU session types. Similar to the control plane, the Inner IP protocol layer is used as the Inner IP of IPSec's "IP in IP" transport mechanism in tunnel mode.

IPSec tunnels have two modes—tunnel mode and transmission mode. The IPSec tunnel used by the UE to access the 5G network through untrusted non-3GPP access technology is tunnel mode. In the early days of Rel-15, 3GPP would choose transmission mode. In the later stage of Rel-15, it was found that the transmission mode could not support Mobike (a function that quickly updates the IPSec tunnel context, usually used for the UE to update the IPSec tunnel context when switching Wi-Fi access points), so 3GPP changed the transmission mode to tunnel mode.

3.10.3.2.2 Processes related to untrusted non-3GPP access

1. REGISTRATION MANAGEMENT PROCESS

The registration management process includes registration and deregistration.

The registration process performed by a UE over an untrusted non-3GPP Access Network is similar to that performed over a 3GPP Access Network. However, differences in Access Networks result in two sets of processes using different mechanisms to establish a connection between the UE and the Access Network and to transmit control plane signaling. Figure 3-127 shows the registration process for untrusted non-3GPP access.

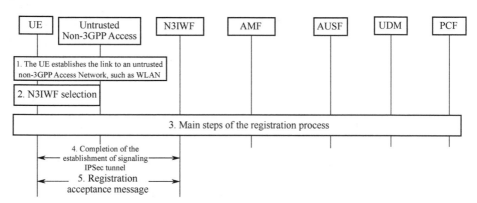

Figure 3-127 Process for registering untrusted non-3GPP access

Step 1: The UE establishes a physical link to an untrusted non-3GPP Access Network (such as WLAN).

Step 2: The UE performs N3IWF selection. See section 6.3.6 of 3GPP TS 23.501 [1].

Step 3: This step is the main step of the registration process. See step 1 to step 6 in section 3.2.1.5.1 of this book. It should be noted that an EAP-5G session will be established between the UE and N3IWF during this process. The AS parameters and NAS signaling used in this step will be transmitted between the UE and N3IWF via an EAP-5G session.

Step 4: Before registration, the UE and N3IWF will complete the process of establishing a signaling IPSec tunnel. After the signaling IPSec tunnel is established, subsequent NAS signaling is transmitted between the UE and N3IWF through the signaling IPSec tunnel.

Step 5: N3IWF completes the registration process by sending the UE a registration acceptance message from the AMF.

See section 4.12.2 of 3GPP TS 23.502 [2] for detailed steps in this process.

The deregistration process performed by the UE over an untrusted non-3GPP Access Network is similar to that performed over a 3GPP Access Network. However, the difference of Access Network leads to differences in the transmission control plane signaling on the Access Network side and resource processing of the Access Network between the two sets of processes:

(1) Transmission control plane signaling on the Access Network side: in the deregistration process performed by the UE through untrusted non-3GPP Access Network, the NAS signaling transmission between the UE and N3IWF needs to be carried by signaling IPSec tunnel; in the deregistration process performed by the UE over the 3GPP Access Network, the NAS signaling transmitted between the UE and the NG-RAN is carried by an air interface RRC connection.

(2) Access Network resource processing: the Access Network resources to be deleted by the UE through the deregistration process performed by untrusted non-3GPP access are the IPSec tunnels (including signaling IPSec tunnels and/or IPSec tunnels used to transport user-plane data) and EAP-5G sessions between the UE and the N3IWF; while the deregistration process performed by the UE through 3GPP access requires the deletion of Access Network resources are the air interface connection resources between the UE and the NG-RAN.

See section 4.12.3 of 3GPP TS 23.502 [2] for detailed steps in this process.

2. Processes related to session management

The session management processes include the session establishment process, session modification process, session release process, and selective deactivation process.

The session management-related processes performed by a UE over an untrusted non-3GPP Access Network are similar to those performed over a 3GPP Access Network. However, due to the difference of Access Network, the two sets of processes differ in the transmission control plane signaling on the Access Network side and resource processing of the Access Network between the two sets of processes:

(1) Transmission control plane signaling on the Access Network side: in the deregistration process performed by the UE through untrusted non-3GPP Access Network, the NAS signaling transmission between the UE and N3IWF needs to be carried by signaling IPSec tunnel; in the session management process performed by the UE over the 3GPP Access Network, the NAS signaling transmitted between the UE and the NG-RAN is carried by the air interface RRC connection.

(2) Access Network resource processing: when the UE performs session management-related processes through untrusted non-3GPP access, the Access Network resources to be established/modified/deleted/deactivated are the IPSec tunnel between the UE and N3IWF; when the UE performs the session management process through 3GPP access, the Access Network resources that need to be established/modified/deleted/deactivated are the air interface connection resources between the UE and the NG-RAN. It should be noted that the IPSec tunnel here is used to transmit user plane data, which is different from the signaling IPSec tunnel established in the earlier registration process.

Regarding QoS issues in processes related to session management, note that the QoS of untrusted non-3GPP on the Access Network side is implemented through IPSec tunnels; that is, the UE and N3IWF transmit different QoS services through different IPSec tunnels. The mapping between the IPSec tunnel on the Access Network and the QoS service flow on the Core Network is one-to-many, which means that an IPSec tunnel on the Access Network side can be used to transmit one or more QoS traffic on the Core Network side. This is consistent with the mapping between DRB of the UE access through 3GPP and QoS service flow on the Core Network side.

For the session establishment process performed by the UE through untrusted non-3GPP access, see section 4.12.5 of 3GPP TS 23.502 [2]; for the session modification process performed by the UE through untrusted non-3GPP access, see section 4.12.6 of 3GPP TS 23.502 [2]; for the session release process performed by the UE through untrusted non-3GPP access, see section 4.12.7 of 3GPP TS 23.502 [2]; see section 4.12.4.3 of 3GPP TS 23.502 [2] for the selective deactivation session flow performed by the Core Network through untrusted non-3GPP access.

3. SERVICE REQUEST PROCESS

The service request process performed by a UE through untrusted non-3GPP access is similar to that performed through 3GPP access. However, the difference of Access Network leads to the differences between the two sets of processes in the triggering conditions of service request process, transmission control plane signaling on the Access Network side, and resource processing of the Access Network:

(1) Triggering conditions of service request process: since non-3GPP access has no paging function, the service request process initiated by the UE through non-3GPP access is not used to respond to paging. Non-3GPP access here includes untrusted non-3GPP access, trusted non-3GPP access, and wireline access.

(2) Signaling on the transmission control plane of the Access Network side: if the UE initiates a service request process in an idle state, in this service request process, AS parameters and NAS signaling need to be transmitted between the UE and N3IWF through the EAP-5G session borne by the IKEV2 protocol; if the UE initiates a service request process in the connection state, the UE and N3IWF need to transmit the NAS signaling through the signaling IPSec tunnel during the service request process.

(3) Access Network resource processing: when the UE performs the service request process through untrusted non-3GPP access, the Access Network resource to be recovered is the IPSec tunnel between the UE and N3IWF (including signaling IPSec tunnel and/or IPSec tunnel for transmitting user plane data); while the Access Network resources that needs to be recovered in the service request process that UE performs through 3GPP access are the air interface connection resources between the UE and the NG-RAN.

See section 4.12.4.1 of 3GPP TS 23.502 [2] for detailed steps in this process.

4. ACCESS NETWORK RELEASE PROCESS

The Access Network release process performed by a UE through untrusted non-3GPP access is similar to that performed through 3GPP access. Nevertheless, the difference in Access Network leads to the difference between the two processes in the processing of Access Network resources.

In Access Network resource processing, when the UE releases Access Network resources through untrusted non-3GPP access, the Access Network resources to be released are the IPSec tunnel (including signaling IPSec tunnel and/or IPSec tunnel for transmitting user plane data) between the UE and N3IWF. However, in the Access Network release process performed by the UE through 3GPP access, the Access Network resources to be released are the air interface connection resources between the UE and the NG-RAN.

See section 4.12.4.2 of 3GPP TS 23.502 [2] for detailed steps in this process.

3.10.3.3 Trusted non-3GPP access

3.10.3.3.1 Trusted non-3GPP access architecture

The architecture of the UE accessing the 5G network through trusted non-3GPP access technology is shown in figure 3-128.

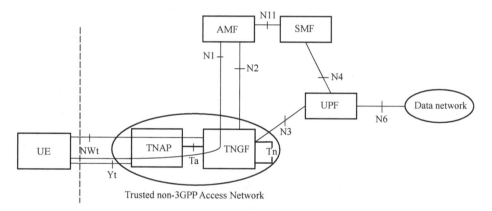

Figure 3-128 Architecture of trusted non-3GPP access technology accessing 5G network

To explain further, the trusted non-3GPP Access Network is a non-3GPP access technology trusted by the operator, such as the non-3GPP Access Network deployed by the operator itself. It consists of two parts: trusted Non-3GPP Access Point (TNAP) and trusted Non-3GPP Gateway (TNGF). TNAP is the signal access point of the UE and the first hop node for the UE to access the 5G network. TNGF, whose network topology location is equivalent to the NG-RAN when the UE accesses through 3GPP, is a trusted non-3GPP access gateway deployed by the operator. It supports N2 and N3 interfaces in 5G Core Network and can deliver NAS signaling between the UE and AMF.

The control plane protocol stack of trusted non-3GPP access before and after the establishment of signaling IPSec tunnel is shown in figures 3-129 and 3-130.

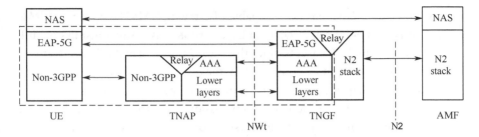

Figure 3-129 Control plane protocol stack for trusted non-3GPP access before signaling IPSec tunnel establishment

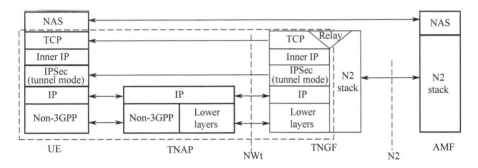

Figure 3-130 Control plane protocol stack for trusted non-3GPP access after the establishment of signaling IPSec tunnel establishment

Compared with the control plane protocol stack of untrusted non-3GPP access, the Core Network control plane protocol stack and NAS protocol layer are consistent with the Core Network control plane protocol stack and NAS protocol layer of untrusted non-3GPP access. The difference mainly lies in the protocol stack on the Access Network side, as shown in the dotted box of figures 3-129 and 3-130.

(1) In figure 3-129, the EAP-5G session, which is established in the registration process and the service request process initiated after the UE enters the idle state, is directly carried on the Layer 2 protocol, such as the Point-to-Point Protocol (PPP) or Ethernet Protocol. However, an EAP-5G session accessed with an untrusted non-3GPP is hosted on the IKEv2 message of Layer 3. The reason for this difference is that the transmission link of the untrusted non-3GPP Access Network is unknown and uncontrollable. It can be Ethernet protocol, Asynchronous Transmission Mode (ATM) protocol or others, so it cannot be guaranteed that EAP-5G is able to transmit on this unknown transmission link. For example, the EAP-5G cannot be transmitted directly over the ATM protocol. Therefore, it is necessary to add a general protocol

layer (i.e., the IKEv2 protocol) capable of carrying EAP-5G to transmit EAP-5G, so as to shield the differences in Layer 2 protocols of the transmission link. For trusted non-3GPP access, since the Layer 2 protocol of the Access Network transmission link can be controlled by the operator, the operator only needs to deploy the Layer 2 protocol that can directly carry EAP-5G as the Layer 2 protocol of the Access Network transmission link. Therefore, when the UE accesses the 5G network through trusted non-3GPP access technology, it can directly transmit an EAP-5G protocol through Layer 2 protocol.

(2) In figure 3-130, the trusted non-3GPP Access Network is trusted by the operator, so the signaling IPSec tunnel established through it does not need encryption; it only performs integrity protection. In comparison, the untrusted non-3GPP Access Network is not trusted by the operator, so the IPSec tunnel established through it needs to be encrypted. It should be noted that although the IPSec tunnel established through trusted non-3GPP does not need to be encrypted, the link between the UE and TNAP still needs to be encrypted to avoid security risks.

The user plane protocol stack for trusted non-3GPP access is shown in figure 3-131.

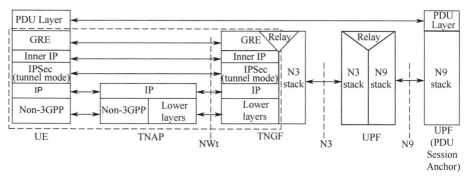

Figure 3-131 User plane protocol stack for trusted non-3GPP access

Compared with the user plane protocol stack of untrusted non-3GPP access, the user plane protocol stack of the Core Network is consistent with the user plane protocol stack of the Core Network of untrusted non-3GPP access. The difference mainly lies in the protocol stack on the Access Network side, as shown in the dotted box in figure 3-131.

Similar to the signaling IPSec tunnel in control plane, the user plane IPSec tunnel established through trusted non-3GPP access does not need encryption, but only performs integrity protection. The user plane IPSec tunnel established through untrusted non-3GPP access needs to be encrypted.

3.10.3.3.2 Related processes of trusted non-3GPP access

1. PROCESSES RELATED TO REGISTRATION AND MANAGEMENT

Processes related to registration and management include registration and deregistration.

The registration process performed by the UE through trusted non-3GPP access is similar to that performed by the UE through untrusted non-3GPP access. However, the differences of Access Networks lead to differences in Access Network security and control plane signaling in Access Network side transmission between the two sets of processes:

(1) Access Network security: the IPSec tunnel established by the UE through trusted non-3GPP access is not encrypted, while the IPSec tunnel established by the UE through untrusted non-3GPP access is encrypted.

(2) Signaling of the transmission control plane in Access Network side: an EAP-5G session of trusted non-3GPP access is directly carried on Layer 2 protocol, while an EAP-5G session of untrusted non-3GPP access is carried on the IKEv2 message of Layer 3.

For the specific steps of this process, see section 4.12a.2 of 3GPP TS 23.502 [2].

The deregistration process performed by the UE through trusted non-3GPP access is basically the same as that performed by the UE through untrusted non-3GPP access. For the specific steps of this process, see section 4.12a.3 of 3GPP TS 23.502 [2].

2. SESSION MANAGEMENT-RELATED PROCESSES

Session management-related processes include the session establishment process, session modification process, session release process, and selective deactivation process.

The process related to session management executed by the UE through trusted non-3GPP access is basically the same as that executed by the UE through untrusted non-3GPP access. Specifically, for the session establishment process performed by the UE through trusted non-3GPP access, see section 4.12a.5 of 3GPP TS 23.502 [2]; for the session modification process performed by the UE through trusted non-3GPP access, see section 4.12a.6 of 3GPP TS 23.502 [2]; for the session release process performed by the UE through trusted non-3GPP access, see section 4.12a.7 of 3GPP TS 23.502 [2]; for the selective deactivation session process performed by the Core Network through trusted non-3GPP access, see section 4.12a.4.3 of 3GPP TS 23.502 [2].

3. SERVICE REQUEST PROCESS

The service request process executed by the UE through trusted non-3GPP access is basically the same as that executed by the UE through untrusted non-3GPP access.

For the specific steps of this process, see section 4.12a.4.1 of 3GPP TS 23.502 [2].

4. PROCESS OF ACCESS NETWORK RELEASE

The Access Network release process executed by the UE under trusted non-3GPP access is basically the same as that executed by the UE under untrusted non-3GPP access.

For the specific steps of this process, see section 4.12a.4.2 of 3GPP TS 23.502 [2].

3.10.3.4 Wireline access

In figure 3-121, wireline access is a wireline Access Network deployed by the operator and supporting access to the 5G network. It consists of two parts: wireline Access Network and wireline access gateway function (W-AGF). Wireline Access Network is an Access Network deployed by fixed network, such as DSLAM or optical line terminal (OLT). W-AGF is the gateway function deployed by the operator to access the 5G network. Its network topology position is equivalent to the NG-RAN when the UE accesses through 3GPP. It supports N2 and N3 interfaces with 5G Core Network.

The fixed network terminal can access the 5G network through wireline access technology. Specifically, there are three types of fixed network terminals that need to access the 5G network through wireline access technology:

(1) Home gateway supporting NAS signaling function, i.e., 5G-RG. See section 3.10.3.4.1 of this book for details.
(2) The existing home gateway of a wireline network that does not support NAS signaling function, namely FN-RG. See section 3.10.3.4.2 of this book for details.
(3) A home terminal accessing the 5G network through a home gateway (including 5G-RG and FN-RG), which is called a 5G Capable UE in this book, for example, computers, mobile phones, and IoT devices supporting 5G functionality. See section 3.10.3.4.3 of this book for details.

3.10.3.4.1 5G-RG access to the 5G network
3.10.3.4.1.1 Architecture of 5G-RG access to the 5G network
The architecture of 5G-RG accessing the 5G network through wireline access technology is shown in figure 3-132.

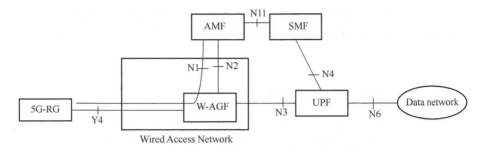

Figure 3-132 The architecture of 5G-RG accessing the 5G network through wireline access technology

The control plane protocol stack of 5G-RG wireline access before and after the establishment of W-CP signaling connection is shown in figures 3-133 and 3-134.

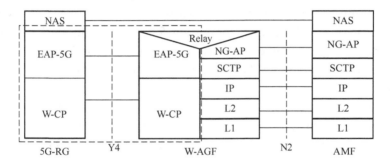

Figure 3-133 The control plane protocol stack of 5G-RG wireline access before the establishment of W-CP signaling connection

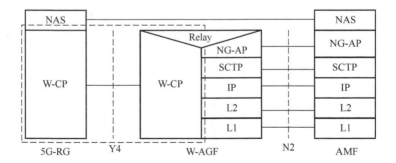

Figure 3-134 The control plane protocol stack of 5G-RG wireline access after the establishment of W-CP signaling connection

Compared with the control plane protocol stack of untrusted non-3GPP access, the Core Network control plane protocol stack and NAS protocol layer are consistent with the Core Network control plane protocol stack and NAS protocol layer of untrusted non-3GPP access. The difference mainly lies in the protocol stack on the Access Network side (as shown in the dotted box of figures 3-133 and 3-134):

(1) In figure 3-133, after the registration process and 5G-RG enter the idle state, the EAP-5G session established in the initiated service request process is hosted on the Wireline access Control Plane protocol (W-CP), while the EAP-5G session of untrusted non-3GPP access is hosted on the IKEv2 message of Layer 3. It is noteworthy that W-CP is not a specific protocol but refers to the wireline access side transmission protocol that transmits EAP-5G session, as parameters and NAS signaling on the wireline access side, such as Point-to-Point Protocol over Ethernet (PPPoE). Since the protocol definition on the wireline access side is not covered by 3GPP, 3GPP refers to the protocol as W-CP.

During the execution of these two processes, 5G-RG and W-AGF will establish a W-CP signaling connection for transmitting NAS signaling to carry NAS signaling in other processes. The function of the W-CP signaling connection is similar to the signaling IPSec tunnel of untrusted non-3GPP access.

(2) In figure 3-134, in other processes (such as deregistration and session management) except for the registration process and the service request process initiated after 5G-RG enters the idle state, NAS signaling will be carried on the W-CP signaling connection and transmitted on the Access Network side.

The user plane protocol stack of 5G-RG wireline access is shown in figure 3-135.

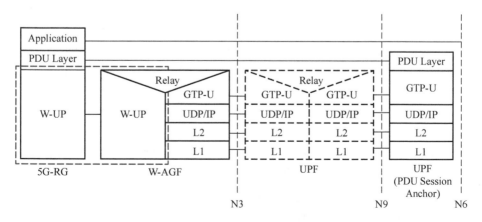

Figure 3-135 The user plane protocol stack of 5G-RG wireline access

Compared with the user plane protocol stack of untrusted non-3GPP access, the user plane protocol stack of the Core Network is consistent with that of the Core Network of untrusted non-3GPP access. The difference mainly lies in the protocol stack on the Access Network side (as shown in the dotted box in figure 3-135).

Similar to the control plane protocol W-CP, 3GPP uses the wireline access user plane protocol (W-UP), which refers to the wireline access side transmission protocol that transmits user plane data on the wireline access side, such as PPPoE protocol.

3.10.3.4.1.2 5G-RG related processes

1. PROCESSES RELATED TO REGISTRATION AND MANAGEMENT

Processes related to registration and management include registration and deregistration.

The registration process performed by 5G-RG through wireline access is similar to that performed by the UE through untrusted non-3GPP access. However, the difference of Access Networks leads to the difference between the two sets of processes on the transmission control plane signaling and access location restriction at the Access Network side:

(1) Signaling of transmission control plane on Access Network side: the EAP-5G session on the wireline access side is hosted on the W-CP protocol, while the EAP-5G session of untrusted non-3GPP access is hosted on the IKEv2 message of Layer 3.

(2) Access location restriction: the service mode of the fixed network is that the fixed network terminal can only obtain services within the location allowed by the contract, such as the user's own home. In order to inherit this service mode, in the process of 5G-RG executing the registration process through wireline access, AMF needs to identify whether the registration process is executed within the location allowed by the contract according to the identification of the line currently connected to 5G-RG sent by W-AGF. AMF allows 5G-RG registration to succeed only when AMF recognizes that 5G-RG is within the location range allowed by the contract.

See section 7.2.1.1 of 3GPP TS 23.316 [20] for the specific steps in this process.

The deregistration process performed by 5G-RG through wireline access is similar to that performed by the UE through untrusted non-3GPP access. However, due to the difference of Access Networks, the two sets of processes have differences in the Access Network side signaling of the transmission control plane and Access Network resource processing:

(1) Signaling of transmission control plane on Access Network side: the NAS signaling on the wireline access side is carried on the W-CP protocol, while the NAS signaling for untrusted non-3GPP access is carried on the signaling IPSec.

(2) Access Network resource processing: the deregistration process of the 5G-RG executed by wireline access requires the release of the W-UP user plane resources and W-CP signaling; in comparison, the deregistration process performed by the UE through untrusted non-3GPP access requires the release of IPSec tunnel user plane resources, signaling IPSec tunnel and EAP-5G session on the Access Network side.

See section 7.2.1.2 of 3GPP TS 23.316 [20] for the specific steps in this process.

2. SESSION MANAGEMENT-RELATED PROCESSES

Session management-related processes include the session establishment process, session modification process, session release process, and selective deactivation process.

The process related to session management performed by 5G-RG through wireline access is similar to that performed by the UE through untrusted non-3GPP access. However, the differences of Access Networks lead to the differences between the two sets of processes in the problems of the Access Network side signaling transmission of the control plane and Access Network resource processing:

(1) Signaling of transmission control plane on Access Network side: the NAS signaling on the wireline access side is carried on the W-CP protocol, while the NAS signaling for untrusted non-3GPP access is carried on the signaling IPSec.

(2) Access Network resource processing: the session management-related processes of the 5G-RG executed by wireline access requires establishing/ modifying/deleting/deactivating the W-UP user plane resources on the wireline Access Network side through executed by wireline access; for the session management-related processes executed by the UE through untrusted non-3GPP access, it is necessary to establish/modify/delete/deactivate the IPSec tunnel user plane resources on the Access Network side.

Specifically, for the session establishment process performed by 5G-RG through wireline access, see section 7.3.1 of 3GPP TS 23.316 [20]; for the session modification process performed by 5G-RG through wireline access, see section 7.3.2 of 3GPP TS 23.316 [20]; for the session release process performed by 5G-RG through wireline access, see section 7.3.3 of 3GPP TS 23.316 [20]; for the selective deactivation session process performed by the Core Network through wireline access, see section 7.3.5 of 3GPP TS 23.316 [20].

3. SERVICE REQUEST PROCESS

The service request process executed by the 5G-RG through wireline access is similar to that executed by the UE through untrusted non-3GPP access. However, the difference of Access Network leads to a difference between the two sets of processes, and there are differences in the Access Network side signaling of the transmission control plane and Access Network resource processing:

(1) Signaling of transmission control plane on Access Network side: if the 5G-RG initiates the service request process in the idle state, in this service request process, AS parameters and NAS signaling need to be transmitted between the 5G-RG and W-AGF through an EAP-5G session carried by the W-CP protocol; if the 5G-RG initiates a service request process in the connection state, in the service request process, the 5G-RG and W-AGF need to transmit NAS signaling through the W-CP signaling connection.

(2) Access Network resource processing: the service request process executed by the 5G-RG through wireline access needs to restore the W-CP signaling connection and/or W-UP user plane resources on the wireline Access Network side; for the service request process executed by the UE through untrusted non-3GPP access, it is necessary to restore the IPSec tunnel on the Access Network side (including signaling IPSec tunnel and/or IPSec tunnel for transmitting user plane data). See section 7.2.2.1 of 3GPP TS 23.316 [20] for the specific steps in this process.

4. ACCESS NETWORK RELEASE PROCESS

The Access Network release process performed by the 5G-RG through wireline access is similar to that performed by the UE through untrusted non-3GPP access. However, the difference in Access Network leads to the difference between the two sets of processes, and there are differences in Access Network side signaling of transmission control plane and Access Network resource processing:

(1) Signaling of the transmission control plane on the Access Network side: the 5G-RG wireline access transmits the NAS signaling of the process through W-CP signaling connection, while untrusted non-3GPP access transmits the NAS signaling of the process through signaling IPSec tunnel.

(2) Access Network resource processing: the Access Network release process executed by the 5G-RG through wireline access needs to restore the W-CP

signaling connection and/or W-UP user plane resources on the wireline Access Network side; for the service request process executed by the UE through untrusted non-3GPP access, it is necessary to restore the IPSec tunnel on the Access Network side (including signaling IPSec tunnel and/or IPSec tunnel for transmitting user plane data).

See section 7.2.5 of 3GPP TS 23.316 [20] for the specific steps in this process.

3.10.3.4.2 FN-RG access to the 5G network
3.10.3.4.2.1 The architecture of FN-RG access to the 5G network

The architecture of FN-RG accessing the 5G network through wireline access technology is shown in figure 3-136.

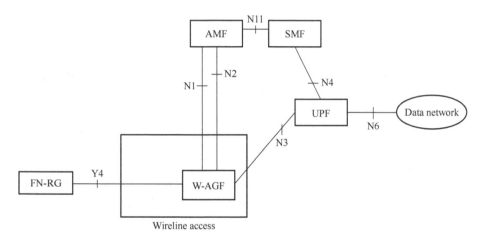

Figure 3-136 The architecture of FN-RG accessing the 5G network through wireline access technology

The key problem for the FN-RG accessing the 5G network is that the FN-RG is an existing home gateway that does not support NAS signaling and supporting NAS signaling is necessary to access the 5G network. Therefore, a network element is needed to help the FN-RG generate and process NAS signaling. This NE is called W-AGF.

That is, the FN-RG only needs to complete the authentication and user plane resource establishment on the wireline access side (between the FN-RG and W-AGF). W-AGF will replace the FN-RG to the 5G Core Network to perform registration management, session management, and other processes, and map the signaling message/user plane connection between the wireline access side and the 5G Core Network side (between W-AGF and AMF/UPF).

The control plane protocol stack of FN-RG wireline access is shown in figure 3-137.

Compared with the control plane protocol stack of the 5G-RG through wireline access, its Core Network control plane protocol stack is consistent with the Core Network control plane protocol stack of untrusted non-3GPP access. The main difference lies in the protocol stack on the Access Network side and the NAS protocol layer of W-AGF (as shown in the dotted box in figure 3-137):

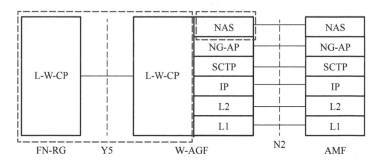

Figure 3-137 The control plane protocol stack of FN-RG wireline access

(1) FN-RG and W-AGF do not need to support the Access Network control protocol layer of mobile networks such as NAS/EAP-5G, as the existing wireline access side protocol can be used. 3GPP uses the name of the Legacy Wireline access Control Plane protocol (L-W-CP) to refer to the wireline access side transmission protocol that transmits control plane signaling on the wireline access side, such as PPPoE protocol.

(2) W-AGF acts as the endpoint of NAS protocol layer and performs the generation and processing of corresponding NAS signaling.

The user plane protocol stack of FN-RG wireline access is shown in figure 3-138.

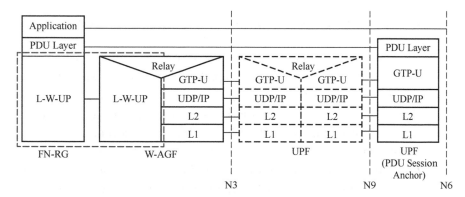

Figure 3-138 The user plane protocol stack of FN-RG wireline access

Compared with the user plane protocol stack of the 5G-RG through wireline access, the user plane protocol stack of its Core Network is consistent with that of the 5G-RG through wireline access. The main difference lies in the protocol stack on the Access Network side (as shown in the dotted box in figure 3-138).

The user plane protocol layer of the Access Network between the FN-RG and W-AGF does not need to support IPSec and other mobile networks, and the existing wireline access side protocol can be used. Similar to the control plane protocol L-W-CP, 3GPP uses the Legacy Wireline access User Plane protocol (L-W-UP) to refer to the wireline access side transmission protocol that transmits user plane data on the wireline access side, such as PPPoE protocol.

3.10.3.4.2.2 FN-RG related processes

1. PROCESSES RELATED TO REGISTRATION AND MANAGEMENT

Processes related to registration and management include registration and deregistration.

The registration process is the process of W-AGF replacing the FN-RG to register to the 5G network. The registration process executed by the FN-RG through wireline access is shown in figure 3-139.

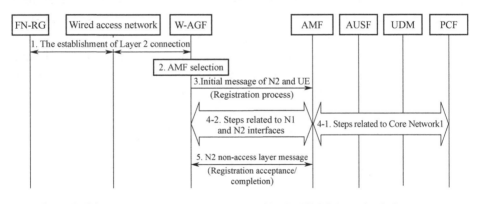

Figure 3-139 The registration process executed by the FN-RG through wireline access

Step 1: FN-RG establishes a Layer 2 connection with W-AGF to access the wireline Access Network. Optionally, in this step, W-AGF will perform authentication on FN-RG, for example, user name/password-based authentication.

Step 2: After the Layer 2 connection between the FN-RG and W-AGF is successfully established, W-AGF will begin the AMF selection.

Step 3: W-AGF takes the place of the FN-RG to send a registration request message to the AMF. It should be noted that W-AGF will also carry an authentication success indication in the N2 message, which contains the registration request message. The authentication success indication is used to notify the AMF that the FN-RG has successfully authenticated in the fixed network.

Step 4: This step is the main part of the registration process, in which step 4-1 refers to the Core Network-related steps in the registration process, and step 4-2 refers to the N1 and N2 signaling related steps of the interaction between the AMF and W-AGF in the registration process.

Step 4-1 consists of authentication, contract acquisition, and scheme acquisition. It should be noted that for the authentication process, 5G Core Network and fixed network use the mechanism of mutual trust. That is, if AMF receives the authentication success indication sent by W-AGF in step 3, the 5G Core Network will not really authenticate FN-RG, but directly feed the authentication success back.

Step 4-2 contains steps related to N1 and N2 interface messages for the interaction between AMF and W-AGF. It should be noted that W-AGF will replace the FN-RG to generate and receive N1 interface messages, i.e., NAS signaling.

Step 5: After successful registration, the AMF sends an N2 interface message to W-AGF, including a registration acceptance message. Accordingly, W-AGF sends a registration acceptance completion message to the AMF instead of the FN-RG, so as to complete the registration process.

See section 7.2.1.3 of 3GPP TS 23.316 [20] for the specific steps in this process.

The deregistration process performed by the FN-RG through wireline access is similar to that performed by the 5G-RG through wireline access. However, different terminal types lead to differences in NAS signaling processing between the two sets of processes: in the deregistration process executed by the FN-RG through wireline access, W-AGF will replace the FN-RG to generate and process NAS signaling; in the deregistration process performed by the 5G-RG through wireline access, the 5G-RG can generate and process NAS signaling by itself.

See section 7.2.1 of 3GPP TS 23.316 [20] for the specific steps in this process.

2. SESSION MANAGEMENT-RELATED PROCESSES

Session management-related processes include session establishment, session modification, session release, and selective deactivation.

In the session establishment process, W-AGF replaces the FN-RG to establish a session in the 5G network. The session establishment process of the FN-RG is shown in figure 3-140.

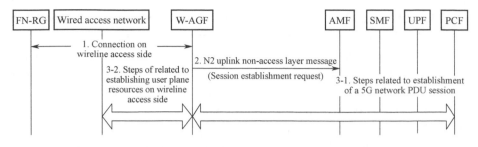

Figure 3-140 The FN-RG session establishment process

Step 1: The FN-RG and W-AGF have established a wireline access side connection during the registration process.

Step 2: W-AGF replaces the FN-RG to send a session establishment request message to the AMF. After the FN-RG successfully performs the registration process through wireline access, W-AGF may perform this step.

Step 3: This step is the main part of the session establishment process. Step 3-1 refers to the relevant steps of establishing a 5G network PDU session, which is consistent with the process of establishing a PDU session by UE; step 3-2 refers to the relevant steps of establishing user plane resources on the wireline access side.

See section 7.3.4 of 3GPP TS 23.316 [20] for the specific steps in this process.

The session modification/deletion/selective deactivation process performed by the FN-RG through wireline access is similar to that performed by the 5G-RG through wireline access. However, the differences in terminal types lead to differences in NAS signaling processing between the two sets of processes: in the session modification/deletion/selective deactivation process executed by the FN-RG through wireline access, W-AGF will replace the FN-RG to generate and process NAS signaling; in the session modification/deletion/selective deactivation process performed by the 5G-RG through wireline access, 5G-RG can generate and process NAS signaling by itself.

Specifically, for the session modification process performed by the FN-RG through wireline access and the session release process performed by the FN-RG through wireline access, see section 7.3 of 3GPP TS 23.316 [20]; for the selective deactivation session process performed by the FN-RG through wireline access, see section 7.3.5 of 3GPP TS 23.316 [20].

3. SERVICE REQUEST PROCESS

The service request process is a process in which W-AGF replaces the FN-RG to perform service requests in the 5G network. The service request process of the FN-RG is shown in figure 3-141.

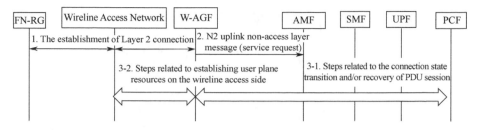

Figure 3-141 The service request process of the FN-RG

Step 1: The FN-RG establishes a Layer 2 connection with W-AGF to access the wireline Access Network. Alternatively, in this step, W-AGF will perform authentication on the FN-RG, for example, username/password-based authentication.

Step 2: W-AGF replaces the FN-RG to send a service request message to the AMF.

Step 3: This step is the main part of the service request process. Step 3-1 refers to the relevant steps of FN-RG from the idle state to connected state and/or the relevant steps of recovering PDU session, which is consistent with the service request process executed by UE; step 3-2 refers to the relevant steps of establishing user plane resources on the wireline access side.

See section 7.2.2 of 3GPP TS 23.316 [20] for the specific steps in this process.

4. ACCESS NETWORK RELEASE PROCESS

The Access Network release process performed by the FN-RG through wireline access is basically the same as that performed by the 5G-RG through wireline access. See section 7.2.5 of 3GPP TS 23.316 [20] for the specific steps in this process.

3.10.3.4.3 5G-capable UE accessing the 5G network

Figures 3-142 and 3-143 describe the architecture of a 5G-capable UE accessing the 5G network. In order to be compatible with Rel-15/16 UE, a 5G-capable UE accesses the 5G network in two ways: untrusted non-3GPP access and trusted non-3GPP access.

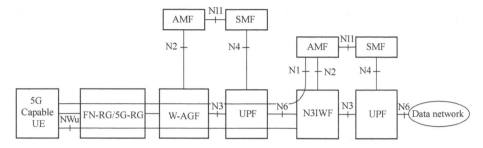

Figure 3-142 5G-capable UE accessing the 5G network through untrusted non-3GPP access technology

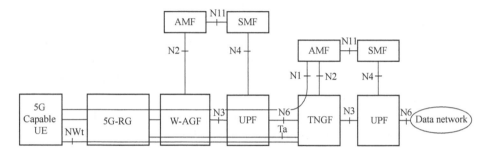

Figure 3-143 5G-capable UE accessing the 5G network through trusted non-3GPP access technology

The related architecture and process of accessing the 5G network through untrusted non-3GPP access technology are consistent with the architecture and process of untrusted non-3GPP access in section 3.10.3.2 of this book, in which:

(1) 5G-capable UE is used as a UE in section 3.10.3.2 of this book, and
(2) FN-RG/5G-RG is accessed as untrusted non-3GPP in section 3.10.3.2 of this book.

The relevant architecture and process of accessing the 5G network through a trusted non-3GPP Access Network are consistent with the architecture and process of trusted non-3GPP access in section 3.10.3.3 of this book, in which:

(1) 5G-capable UE is used as UE in section 3.10.3.3 of this book, and
(2) 5G-RG as TNAP in section 3.10.3.3.

In Rel-16, a 5G-capable UE accesses the 5G network through trusted non-3GPP access technology, which is only applicable to the scenario where a 5G-RG and a 5G-capable UE belong to the same PLMN, while a 5G-capable UE accesses the 5G network through untrusted non-3GPP access technology without this restriction. This is because through trusted non-3GPP access, 5G-RG and TNGF will have AAA interaction. If a 5G-RG and a 5G-capable UE belong to different PLMN, 5G-RG cannot route AAA messages to TNGF.

3.10.3.5 Hybrid access

In figure 3-121, hybrid access refers to the state in which UE or the 5G-RG connects 3GPP Access Network and non-3GPP Access Network at the same time. Hybrid access can not only increase the bandwidth provided by 5G network for the terminal through the shunting function but also maintain service continuity when the terminal loses the coverage of an Access Network (e.g., the terminal moves out of the coverage of an Access Network).

Hybrid access has two modes: single-access PDU session mode and multi-access PDU session mode. Among them, the single-access PDU session mode is the hybrid access mode supported by 3GPP Rel-15, and the multi-access PDU session mode is the hybrid access mode supported by 3GPP Rel-16.

For the UE, the single-access PDU session mode means that the UE establishes independent PDU sessions on the 3GPP Access Network and non-3GPP Access Network (including untrusted non-3GPP access and trusted non-3GPP access) simultaneously and uses respectively independent PDU sessions on both sides for data transmission. The multi-access PDU session mode is that UE establishes a PDU session called a "multi-access PDU session" in the 3GPP Access Network and non-3GPP Access Network (including untrusted non-3GPP access and trusted non-3GPP access). It can use two Access Networks for data transmission at the same time, and aggregate and divert data in UPF.

Similarly, for the 5G-RG, the single-access PDU session mode means that the 5G-RG establishes independent PDU sessions simultaneously in the 3GPP Access Network and wireline access and uses independent PDU sessions on both sides for data transmission. The multi-access PDU session mode is that the UE establishes a PDU session called a "multi-access PDU session" in the 3GPP Access Network and wireline access. The multi-access PDU session can use two Access Networks for data transmission at the same time, and aggregate and divert data in UPF.

The single-access PDU session mode belongs to the coarse-grained hybrid access mode. It supports the diversion of PDU session granularity; that is, the services on different PDU sessions of the same terminal can be transmitted through different Access Networks.

Compared with the single-access PDU session mode, the multi-access PDU session mode supports finer granularity and more flexible service diversion. For example, the multi-access PDU session mode can divert an application and adjust the diversion strategy in real time according to the network state. For the multi-access PDU session mode, see section 5.32 of 3GPP TS 23.501 [1] and section 4.22 of TS 23.502 [2].

3.10.3.6 IPTV

5G fixed-mobile convergence is the convergence of network access and network services. IPTV is precisely such a critical fixed network service. The 5G architecture for supporting IPTV services is shown in figure 3-144, where an IPTV set-top box connects to the IPTV data network through the PDU session data channel of the 5G-RG, for IPTV services.

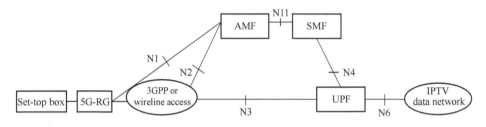

Figure 3-144 The 5G structure supporting IPTV services

In 3GPP Rel-16, the 5G architecture for fixed-mobile convergence supports the 5G-RG in acquiring IPTV services through fixed access mode and wireline access. The scheme can be generally divided into three phases:

(1) The IPTV registration and session establishment process: the 5G-RG initiates the registration process and establishes the session used to acquire IPTV services.
(2) The IPTV access process: this process is used to connect the set-top box to the IPTV network, which includes the completion of the set-top box authentication process by the IPTV network and the allocation process of IP addresses.
(3) The IPTV service transmission process: the process includes the transmission of unicast data and multicast data.

1. The IPTV registration and session establishment process
The registration process for the 5G-RG acquisition of IPTV services is the same as the usual registration process, but the session establishment process requires the following enhancements:

(1) The operator needs to deploy N6 interfaces that can access the IPTV network and configure the corresponding IPTV Data Network Name (DNN). In the PDU session establishment process, the 5G-RG needs to carry the IPTV DNN in the session establishment request message to request the establishment of a PDU session capable of acquiring IPTV services. 5G networks will select the network element (including SMF selection and UPF selection) based on the IPTV DNN.

(2) In the PDU session establishment process, the 5G-RG requests the IP address through DHCP to be compatible with the IPTV network and the existing network set-top box. This is because the existing network set-top box obtains the IP address from the IPTV network through DHCP after successfully connecting to the IPTV network.

(3) Mobile network operators need to collaborate with IPTV operators to configure a Line ID for the 5G-RG in the contract, which will be sent to the SMF as part of the contract data during the PDU session establishment process. This is because the IPTV network will authenticate the set-top box based on the Line ID carried in the DHCP message sent by the set-top box. To be compatible with the existing IPTV network, the DHCP message sent by the set-top box over the 5G network should also carry this Line ID. It should be noted that the Line ID generally refers to the Line ID of the cable network, but here the Line ID will not authentically identify the fixed network line but only be used for authentication during IPTV access. This step prepares for the execution of the IPTV access process in Stage 2.

(4) The PCF will obtain the multicast access control list from the IPTV network through the Capability Open Function Element and send the multicast access control information carried by the multicast access control list to the SMF and UPF during the session establishment process. This is done because the current fixed network will control the multicast authorization of the IPTV service based on the multicast access control list. For example, the multicast access control list will describe whether a user is fully/partially/disallowed to preview a certain multicast TV channel. To be compatible with the existing multicast permission control function, the 5G network also needs to perform multicast authorization control based on the multicast access control information carried by the multicast access control list. This step is the preparation to execute the IPTV service transmission process in Stage 3.

2. THE IPTV ACCESS PROCESS

The process of a set-top box accessing IPTV through the 5G network is shown in figure 3-145.

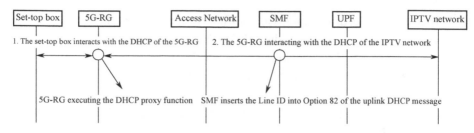

Figure 3-145 The set-top box accessing IPTV through the 5G network

This process consists of two DHCP interaction steps that are executed in a nested fashion. During the nested execution of these two DHCP interaction steps, the 5G-RG serves the DHCP proxy function. In other words, the 5G-RG plays the role of DHCP server for the set-top box and the role of DHCP client for the IPTV network.

Step 1: The process of performing DHCP message interaction between the set-top box and the 5G-RG.

Step 2: The 5G-RG performs the process of DHCP message interaction with the IPTV network. In this process, when the SMF receives an uplink DHCP message from the 5G-RG, the SMF needs to insert the line identifier obtained in stage 1 into the Option 82 of the uplink DHCP message, and then send the DHCP message after inserting the Line ID to the IPTV network. The IPTV network needs to authenticate the STB based on the Line ID received from the DHCP message. If the authentication is successful, the IPTV network will assign an IP address through the downlink DHCP response message.

At this point, the set-top box has completed the process of accessing the IPTV network through the 5G network and can now obtain IPTV services.

3. THE IPTV SERVICE TRANSMISSION PROCESS

The 5G network naturally supports the transmission of unicast data but cannot support the transmission of multicast data. The reason is that the destination address carried by downlink multicast data is a multicast address. When the UPF receives downlink multicast data, it cannot identify the corresponding PDU session based on the multicast address, so it cannot complete the transmission of multicast data.

The way in which 5G supports IPTV multicast data transmission is shown in figure 3-146.

The transmission process of IPTV multicast data is as followed:

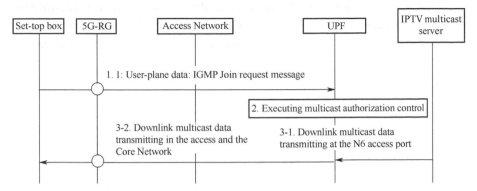

Figure 3-146 5G supporting IPTV multicast data transmission

Step 1: The set-top box sends a join request of the Internet Group Management Protocol (IGMP Join) to the 5G-RG, requesting to get the service of an IPTV multicast channel. 5G-RG performs Network Address Translation (NAT) and sends the IGMP Join message to the UPF via PDU session.

Step 2: The UPF performs multicast authorization control based on the multicast access control information; that is, the UPF determines whether the PDU session is allowed to obtain the requested multicast data based on the multicast access control information. If allowed, the UPF will add the PDU session to the transmission list corresponding to the multicast channel.

Step 3: When UPF receives the downlink multicast data, it will transmit the downlink multicast data to the 5G-RG through the PDU session recorded in the multicast channel transmission list, and then the 5G-RG will transmit it to the set-top box, thus completing the transmission of multicast data.

3.10.4 Summary

Content about 5G fixed-mobile convergence consists of four main components:

(1) Fixed wireless access, which realizes the function of fixed network terminals to access 5G networks through 3GPP access technology

(2) Non-3GPP access, which enables mobile terminals and fixed network terminals to access 5G networks through non-3GPP access technologies

(3) Mixed access, realizing the function of the mobile terminal and fixed network terminal accessing the 5G network through 3GPP and non-3GPP access technologies at the same time

(4) IPTV service support, realizing the function of fixed network terminals to access IPTV service through the 5G network

With these four functions, the 5G fixed-mobile convergence feature enhances network coverage, simplifies integrated network operation and maintenance, improves network bandwidth, and provides access-independent convergence services to terminal users.

3.11 Security

3.11.1 Introduction

5G faces both opportunities and challenges arising from new services, architectures, and technologies, as well as higher user privacy protection requirements. During the development of 3GPP's SA3 working group security standards, technical experts from more than 70 companies around the world analyzed security threats and risks in several areas, including security architecture, access authentication, security context, and key management, RAN (radio Access Network) security, authorization, user contract information privacy protection, network slicing security, network domain security, security visualization, and security configuration management, credential distribution, 4G network and 5G network interoperation security, IoT security, broadcast/multicast security, management surface security, and risk analysis of cryptographic algorithms.

The goal of 5G network security is the same as that of 4G, which is to ensure the confidentiality, integrity, and availability of the network and data. With 5G networks facing new potential security challenges such as technological changes over the next ten years, 5G security technologies need to be enhanced and improved from those used for 4G.

3.11.2 5G security architecture

The 5G mobile network designs its security architecture according to the principle of hierarchical sub-domains. The overall framework of 5G security is shown in figure 3-147, which is divided into the following six security aspects:

(1) Network Access Security (I): a set of security features that enable the UE to securely pass through the network for authentication and service access, including 3GPP access and non-3GPP access, and protect the security of each interface (especially the air interface). It includes the transmission of security contexts from the service network to the Access Network to achieve access security. Specific security mechanisms include bi-directional access authentication, transmission encryption, and integrity protection.

(2) Network Domain Security (II): a set of security characteristics that enable network nodes to exchange signaling data and user-plane data securely.

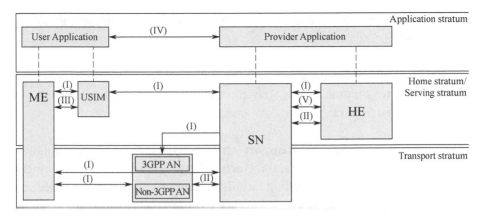

Figure 3-147 Overall 5G security architecture

Network domain security defines the security characteristics of the interface between the Access Network and the Core Network, as well as the security characteristics of the interface from the service network to the attribution network. As with 4G, the 5G access and Core Networks are separated with clear boundaries and the interfaces between the access and Core Networks can be secured using security mechanisms (e.g., IPSec).

(3) User domain security (III): a set of security features that allow users to access a mobile device securely. Security mechanisms (such as PIN codes) are used inside the terminal to ensure security between the mobile device and the USIM card.

(4) Application domain security (IV): a set of security features that enable applications in the user domain (terminal) and applications in the provider domain (application server) to exchange messages securely. The security mechanism of this domain is transparent to the entire mobile network and requires security services by the application provider.

(5) Service-based architecture domain security (V): a set of security features that enable network functions of a service-based architecture to communicate securely within the service network domain and between the service network domain and the attributed network domain. These features include security aspects such as network function registration, discovery and authorization, and the protection of service-based interfaces. This is a new security domain for 5G. 5G Core Networks use a service-based architecture and require appropriate security mechanisms to ensure security between 5G Core Network functions. The main security mechanisms in this domain include TLS (Transport Layer Security) and OAUTH (Open Authorization).

(6) Security visualization and configurability (VI): a set of security features that allows the user to know whether a security feature is enabled or not. It is worth

mentioning that in figure 3-147, security visualization and configurability are not visible.

As with 4G, the 5G security architecture can be further divided into a Transport Stratum, Service Stratum/Home Stratum, and Application Stratum, which are securely isolated from each other:

(1) Transport stratum: the transport stratum guarantees the transmission security of signaling and data interacting between the terminal equipment and network side. Specifically, it includes encryption and integrity protection for signaling and data transmission, as well as interface security between the Access Network and the Core Network.

(2) Service stratum/home stratum: the service stratum/home stratum guarantees the ability of terminal devices to obtain the access service of the service/home operator safely, which specifically includes the authentication of the terminal devices by the network of the home operator, the terminal devices accessing the USIM card safely in response to the authentication service, the interface security between the network elements of the service/home operator, and the security of the service-based architecture between the services.

(3) Application stratum: the application stratum guarantees the security of terminal devices and service providers, which means that it provides end-to-end security protection between terminal devices and application services.

The 5G network inherits the layered and domain security architecture of the 4G network, with clear boundaries between the access and Core Networks, interconnection through standard protocols, support for interoperability between different vendor devices, and a standards-based security protection mechanism. For regulators, network security risks need to be assessed from these three strata.* For service providers, network security risks need to be managed end-to-end from the application stratum. For operators, network security risks need to be managed from the serving/attribution and transport stratum. Equipment providers need to focus on network equipment security risks. The entire industry should work together to address service, architectural, and technical security risks under a standard architecture.

3.11.3 Key features of 5G security

Independent 5G networks support additional security features to address future security challenges that may arise over the 5G lifecycle. Non-independent 5G

---

* The plural 'f 'stratum' is "strata," which is from the Latin. The Latin plural is used here.—Trans.

networks have the same security mechanisms as 4G networks and will continue to improve their security levels through coordinated standards development and practices.

The following are several of the enhanced security features of 5G versus 4G.

3.11.3.1 Integrity protection function on the user plane of the air interface

5G has better air interface security. In 2G/3G/4G, user data encryption protection is implemented between the user and the network, on top of which the 5G standard further supports the integrity protection mechanism for user data to prevent user data tampering attacks. The security risks associated with no integrity protection on the UP side of 4G network air interface are shown in figure 3-148.

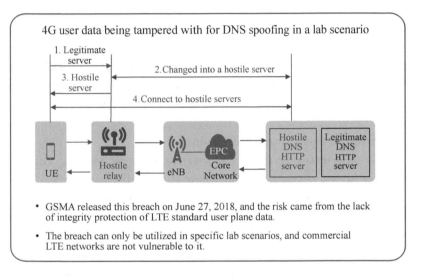

Figure 3-148 The security risks associated with no integrity protection on the UP side of 4G network air interface

Source: https://alter-attack.net/

During the development of the 5G security standard, integrity protection has been added for specific attacks on the user plane of this air interface, to prevent user plane data from being tampered with and improve security. The integrity protection algorithm is derived from the same integrity protection algorithm as the control plane CP.

The addition of integrity protection on the UP side of the 5G network air interface is shown in figure 3-149.

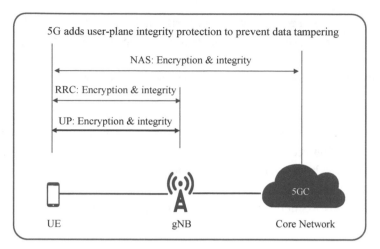

Figure 3-149 The addition of integrity protection on the UP side of the 5G network air interface

At this point, encryption and integrity protection at the NAS layer and encryption and integrity at the RRC and UP plane have been realized in the 5G standard.

3.11.3.2 Enhanced user privacy protection

In 2G/3G/4G, the user's persistent identity IMSI is sent in clear text over the air interface. Attackers can exploit this flaw to track the user. In 5G, the user's permanent identity SUPI will be sent in encrypted form to prevent "IMSI Catcher" attacks.

The clear text transmission of IMSI was a necessary link in mobile standards prior to 5G. An attacker, as depicted in figure 3-150, can attack to capture the IMSI in the air interface transmission when the UE sends an attachment request, and take the opportunity to lock down the location of a specific IMSI user, causing users' privacy leakage.

The 5G network introduces a public key encryption mechanism that enables users to encrypt IMSI information when they first attach to the network to protect IMSI. The SUPI containing IMSI is encrypted by the public key of the attribution network to obtain a one-time SUCI, which is transmitted in the air interface instead of the SUPI, enabling the user's identity information to be protected against IMSI Catcher attacks. The 5G network introduces SUCI, encrypted transmission of IMSI to protect user privacy, as shown in figure 3-151.

3.11.3.3 Better roaming security

The connection between carriers usually needs to be established through a transit carrier. An attacker can impersonate a legitimate Core Network node and launch

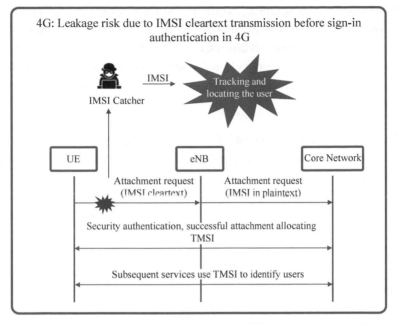

Figure 3-150 The possibility of privacy leakage due to IMSI Catcher attacks in 4G networks

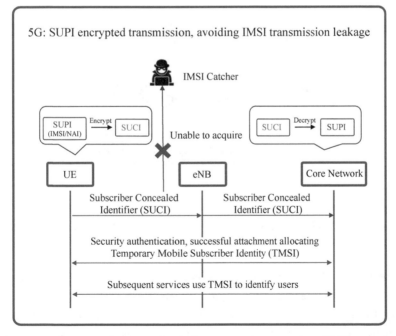

Figure 3-151 5G network introduces SUCI, encrypted transmission of IMSI protecting user privacy

an SS7 attack by controlling a transit operator's device. The current roaming protection mechanism of the 4G network is not perfect, and the roaming security boundary is at risk. Examples of this include (1) no filter to block invalid messages, (2) no end-to-end signaling protection, (3) untrusted or insecure service partners.

TLC is used between 5GC functional modules to protect the security of information transmitted.

(1) Encryption and integrity protection of transmitted data via TLS
(2) Use of TLS bi-directional authentication to prevent fake NFs from accessing the network

For non-serviced architectures, the use of IPSec to protect the security of information transmitted between 3GPP network elements is supported.

(1) Ensuring confidentiality and integrity of data transmission through IPSec encryption and checksum
(2) Authenticity of the data source ensured by IPSec authentication

TLS is used to secure connections between network elements in the 5G Core Network, as shown in figure 3-152.

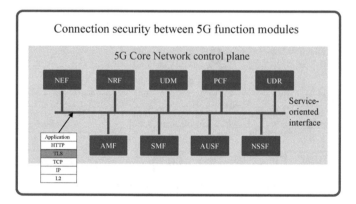

Figure 3-152 5G Core Network elements use TLS to secure connections between network elements

SEPP is defined under the service-oriented architecture in 5G to provide end-to-end security for inter-operator signaling at the transport and application layers, making it impossible for inter-operator transit devices to eavesdrop on sensitive information (e.g., keys, user identities, SMS messages) interacted between Core Networks. 5G roaming security architecture is shown in figure 3-153.

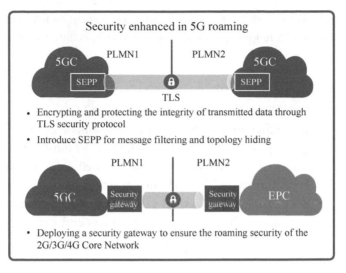

Figure 3-153 5G roaming security architecture

3.11.3.4 Cryptographic algorithm enhancements

The 5G key length can be increased to 256 bits, as shown in figure 3-154. To cope with the potential impact of the emergence of future quantum computers on the cryptographic algorithms in the 3GPP specification, 5G is exploring support for 256-bit algorithms. Meanwhile, 3GPP has recommended that ETSI SAGE evaluate 256-bit algorithms.

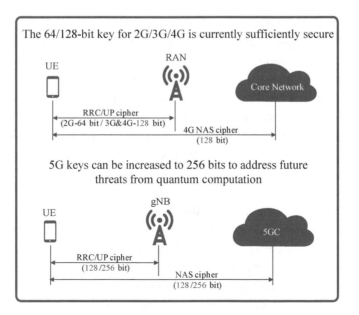

Figure 3-154 5G key length can be increased to 256 bits

5G network security standards include additional security features to address potential security challenges and enable security enhancements over the future 5G lifecycle.

3.11.4 5G security assessment

The GSMA and 3GPP have jointly developed the NESAS (Network Product Security Assurance Scheme) for the security assessment of mobile network devices. The program provides a security baseline to demonstrate that network devices meet a series of security requirements, as well as demonstrating that manufacturers' development and lifecycle management processes meet security standards. The 3GPP has initiated several security assessment standards for 5G network equipment, and mainstream device providers and carriers are actively participating in the GSMA's NESAS standard development efforts.

3.11.4.1 Supervising the testing process for NESAS and SCAS

The general overview of the standard documents of the Network Equipment Security Assurance Scheme (NESAS), shown in figure 3-155, is the overall framework of the standard documents of the NESAS project, in which the standards developed by GSMA focus on supervision and assessment of production processes, while the SCAS series of schemes developed by 3GPP provides specific security assessment baselines for communication devices. NESAS and SCAS each have their focus. The organic combination of the two constitutes the industry's authoritative security assessment specification for communications devices.

GSMA's SECAG working group is mainly responsible for defining the process specification and requirements of the NESAS program, as well as supervising and evaluating the production process of equipment manufacturers. In figure 3-155, the above three documents are developed by the SECAG working group of GSMA. 3GPP's SA3 Security Working Group is mainly responsible for the development of test and evaluation standards for SCAS, and the following three documents in figure 3-155 are developed by them. The standard contains the definition of security requirements and testing cases. The division of labor between GSMA and 3GPP in the Network Product Security Assurance Scheme (NESAS) is shown in figure 3-156.

The overall operation process of the Network Equipment Security Assurance Scheme (NESAS) is shown in figure 3-157. The overall operation of the NESAS as defined by GSMA contains NESAS specification development and application, the appointment of an independent supervising team, supervising report output, the appointment of a qualified security test laboratory, process supervising and test equipment provision by equipment manufacturers, and test standards according to SCAS conducting security evaluation and other standard specifications. The test results are finally measured in two dimensions: the manufacturing process of the equipment provider and the security function test of the tested product. The equipment vendor supervising mechanism of the Network Equipment Security Assurance Scheme (NESAS) is shown in figure 3-158.

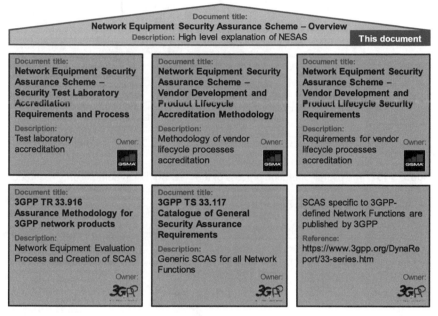

Figure 3-155 Overview of the standard documentation for the Network Equipment Security Assurance Scheme (NESAS) [21]

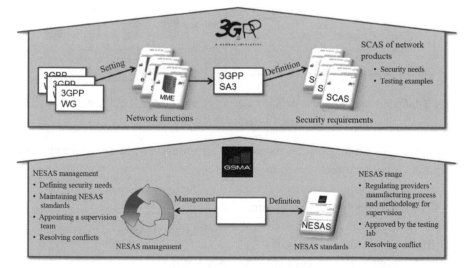

Figure 3-156 The division of labor between GSMA and 3GPP in the Network Equipment Security Assurance Scheme (NESAS) [21]

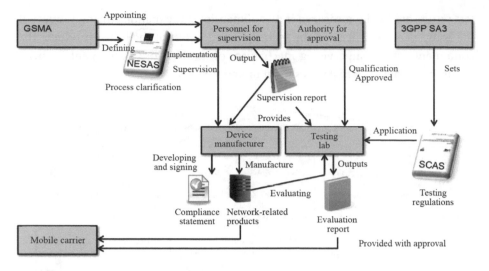

Figure 3-157 The overall operation process of the Network Equipment Security Assurance Scheme (NESAS) [21]

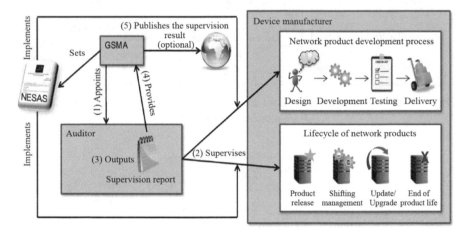

Figure 3-158 Network Equipment Security Assurance Scheme (NESAS) equipment manufacturer supervising mechanism [21]

A more specific process for assessing the production process supervision of equipment manufacturers is as follows:

(1) GSMA appoints supervising team.
(2) Supervising team supervises the production process.
(3) The team produces a report.
(4) Such report is then submitted to the GSMA team.

The specific process for evaluating a network equipment product:

Step 1: The supervision team supervises the production process of the equipment manufacturer and provides a report. In this process, only the specifications developed by the GSMA organization are required.

Step 2: The test lab provides the results.

Step 3: The test lab performs a security assessment of the device. The security assessment process uses the SCAS family of standards defined by 3GPP as a baseline and evaluates the device in conjunction with the specifications developed by the GSMA organization.

Step 4: The test lab writes the assessment report.

3.11.4.2 Summary of security assessment standards for NESAS and SCAS

For equipment suppliers, NESAS and SCAS can benefit them in several ways:

(1) providing certification from the world's leading mobile service representative body
(2) providing world-class security reviews of security-related processes
(3) providing a unified approach to security supervision
(4) avoiding fragmented and potentially conflicting security assurance requirements in different markets

For mobile operators, NESAS and SCAS can benefit in several ways:

(1) setting strict security standards and professional test cases that require a high level of commitment from manufacturers
(2) allowing manufacturers to implement appropriate security measures and practices
(3) eliminating the need for carriers to invest additional money and time in separate vendor audits

3.11.5 Summary

5G is becoming a reality and will continue to evolve. Based on the successes of 4G security, the current risk-controlled state of 5G security is the result of the joint efforts of the whole industry. In the face of future risk control in the 5G lifecycle, it is necessary to continuously enhance security solutions through technological innovation, and to construct secure systems and networks through standards and ecological cooperation.

3.12 Vertical industry-oriented features

3.12.1 5G LAN
3.12.1.1 Introduction
The three typical scenarios for LAN services discussed in the 3GPP standard are

(1) data sharing in a home local area network,
(2) device interconnection in an enterprise office network,
(3) interoperability of industrial controllers and actuators under factory local area networks.

The common feature of the above three scenarios is that they are all local area networks (LAN), i.e., data interoperability between terminals and terminals in a LAN environment, which is different from the traditional "Client-Server" communication mode in 3GPP networks. 5G LAN adds support for LAN services in 5G networks, allowing data interoperability between two peer terminals in 5G networks. In other words, the control plane and user plane of the 5G network are enhanced to support the local exchange of the UE data, the main advantages of which are (1) UE data does not come out of the UPF, which reduces data transmission outside the N6 interface and the processing delay in the application server, and reduces the end-to-end delay; (2) UE data does not go out of the UPF, and the end-to-end data path is all within the scope of 3GPP management, which can better guarantee QoS and increase end-to-end communication reliability. As to the 5G network support for LAN services, an important part of the previous standard is missing—LAN creation and management. Vertical industry users want to fill this gap and allow users to provide LAN services natively over 5G networks.

The 5G LAN topic implements the 5G network's awareness of the LAN, including LAN creation, authorization, and management. For the UEs involved in LAN communication, the 5G network is targeted to configure optimized communication paths for them to ensure that communication data exchange occurs as locally as possible.

3.12.1.2 Structure and major solutions
The 5G LAN enhances the 5G system in three ways:

(1) Network management of LAN resources and LAN membership: the 5G network provides LAN services to vertical industry users and supports dynamic LAN creation, where the user simply provides the LAN membership identification to the network, which stores LAN membership information

and plans LAN communication resources (such as SMFs and UPFs that serve this LAN).

(2) LAN-level session management: when a LAN member (UE) initiates a service to access LAN communication, the network correlates the session context of this UE with the sessions of other LAN members (UE) to construct a LAN-level session context so that the data plane forwarding rules between two or more LAN members can be reasonably formulated.

(3) User plane architecture enhancement to support local switching for LAN traffic: depending on the LAN-level session context, when two or more UEs' sessions are served by the same UPF, a local sparing forwarding mode is used, and the UPF is required to enhance its ability to support local data transfer; if more than one UPF serves group members, then a forwarding tunnel can be established between any two UPFs for LAN forwarding of data services.

The 5G LAN service logic diagram is shown in figure 3-159, which reflects the main features of the 5G LAN.

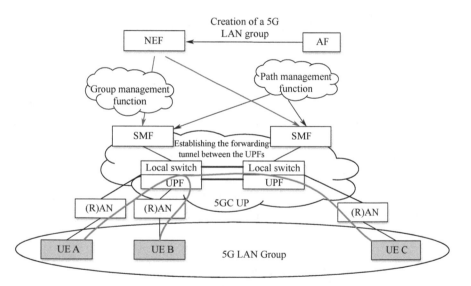

Figure 3-159 5G LAN service logic diagram

3.12.1.3 Summary

The 5G network adds LAN management and LAN communication configuration capabilities in the Rel-16 phase to meet the needs of users accessing LAN services. LAN services are utilized in homes, enterprises, and factories. With these capability enhancements, the 5G network goes further in meeting the applications of vertical industry users.

3.12.2 Time-sensitive transmissions

3.12.2.1 Introduction

During the discussion phase for industrial Internet requirements, 3GPP first clarified the service requirements to support TSN networks. TSN is a mature standard defined by IEEE that is widely used in industry, and for many vertical industry users, they want 5G networks to replace the wireline access of traditional TSN networks. When devices access the TSN network through 5G systems, they are able to implement the existing control functions of the TSN network while meeting various performance metrics of the TSN network data transmission.

The support of 3GPP for TSN services solves these two main issues:

(1) Clock synchronization, i.e., the clock information of the TSN network is synchronized to the UEs accessed through the 3GPP network.
(2) Deterministic transmission, i.e., the 5G system supports sending TSN service messages to the next-hop TSN node through the UE side or UPF side within a determined time zone.

The whole scheme architecturally views the 5G network as a TSN switch, and the UE and UPF as ports at both ends of this TSN switch. The processing capability of TSN messages is added to the UE and UPF to adapt to the user plane transmission of the 5G system.

The clock synchronization scheme does not rely on the base station to sense the external TSN clock, but only requires the UE and UPF to be synchronized with the TSN clock. The deterministic transmission scheme reports the end-to-end transmission delay of the 5G network to the TSN controller, which reinjects the transmission requirements of TSN messages into the 5G network, and the SMF serves as the control point of each node in the use plane of the 5G network, further decomposing and implementing the requirements of the TSN controller to the base stations and UPFs.

3.12.2.2 Architecture and basic solution

Looking at the entire TSN network, the 5G system is one of the TSN bridges, whose logical architecture is shown in figure 3-160.

Solutions for time synchronization are as follows:

(1) Basic idea: the devices, UEs, base stations (gNB) and UPFs in the 5G system are first synchronized to the 5G system clock. TSN network clock information is carried in generalized Precision Time Protocol (gPTP) messages, which are transmitted via NW-TT (Network TSN Translator, a function of UPF) to DS-TT (Device-side TSN Translator, which can be a function integrated

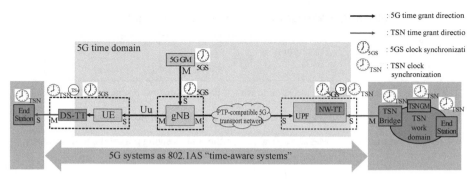

Figure 3-160 5G system as a TSN bridge in a TSN network

into the UE). The UE calculates the deviation of the 5G clock from the TSN clock based on the timestamp information in the gPTP message and gets the accurate TSN clock information.

(2) 5G clock synchronization method: information is being broadcast by the base station to the UE by way of air interface frame alignment, and the UE is able to calculate the accurate local 5G clock information; the UPF can calculate the accurate local 5G clock information through the 5G clock information provided by the base station or the transmission network.

(3) TSN clock delivery method: the TSN GM sends clock synchronization messages to each TSN node, TSN Bridge or End Station. Since the 5G system is also a TSN Bridge, it will also receive clock synchronization messages. Specifically, NW-TT is the first module in the 5G system that receives the TSN clock information (carried in the gPTP message). It adds the time stamp of the receipt of this message in the gPTP message header and encapsulates the gPTP message in the GTP message on the 5G user side. The UPF co-created with NW-TT sends the encapsulated message to the UE and further to DS-TT (which can be co-located with the UE) according to the regular user plane message to restore the TSN clock information in the gPTP message and to calculate the local TSN clock information.

Deterministic transmission solutions are as follows:

(1) 5G system treats data messages in TSN networks as a special kind of service, namely Time Sensitive Communication (TSC), and designs new QoS parameters for data transmission of this service, called Time Sensitive Communication Assistance Information (TSCAI). See table 3-35.

Table 3-35 Information supporting time sensitive communication

Supporting information content	Description
The direction of data flow	The direction in which the TSC data flow is sent, uplink or downlink
Cycle	The time interval between the arrival of two messages
Message arrival time	Message arrival time, either the receiving port of the base station (downlink messages) or the sending port of the UE (uplink messages)

The TSCAI is generated by the SMF and provided to the base station based on the control requirements of the external TSN controller. Based on this information, the base station determines the scheduling method for TSC messages, such as the reservation of air interface resources through semi-static scheduling.

(2) The deterministic feature of TSC messages is reflected in the fact that messages arrive at a TSN node device neither too early nor too late, i.e., there is a time window for the arrival of messages. However, the transmission principle of user-plane devices in 5G networks is usually first-come-first-served, and by caching messages on DS-TT and NW-TT, the purpose is to correct the message delivery time to ensure that messages are sent within the time window, thus also ensuring that the message arrives at the next-hop TSN node within a time window.

3.12.2.3 Summary
The TSN network is a mature communication model for industrial applications. To allow devices in industrial environments to access TSN services through 5G networks and communicate with devices in other TSN networks, the 5G network needs to be adapted into a TSN switch that can be synchronized with the TSN network clock under the control of a TSN controller and can achieve deterministic transmission.

3.12.3 Network deployment within vertical industries
3.12.3.1 Introduction
In order to support the vertical industry-oriented deployment of 5G systems to facilitate the provision of wireless access and communication services to vertical industry terminals, the 3GPP standard defines the concept of Non-Public Network (NPN) in Rel-16 phase.

The specific technical requirements for 3GPP 5G systems to support NPN are described in 3GPP TS 22.261 [22] and comprise the following three aspects:

(1) Support for providing 5G network coverage for NPNs in a specific geographical location and support for enterprise independent operation of NPNs.
(2) Support for NPN access control, including not allowing the UE to automatically select access PLMN or another NPN when contracting is not allowed; UEs with PLMN contracts only cannot automatically opt-in to NPN.
(3) Support for service interworking between NPN and PLMN, i.e., support a UE to access the PLMN network from NPN and use the PLMN service, and when the UE moves between NPN and PLMN network, ensure that the continuity of the PLMN service is satisfied, and vice versa.

Based on the above requirements, 3GPP defines two NPN deployment modes: Public Network Integrated NPN (PNI-NPN) and Stand-alone NPN (SNPN), and defines the NPN network identification, terminal configuration, and interworking with PLMN services under the two deployment modes respectively. In addition to 3GPP, other standards organizations have discussed vertical industry-specific 5G system deployment scenarios. For example, the 5G Alliance for Connected Industries and Automation (5G-ACIA) has published a white paper on non-public network deployment scenarios, which explores the deployment relationship between non-public and public networks.

3.12.3.2 Deployment scenarios defined by 5G-ACIA

The 5G Alliance for Connected Industries and Automation (5G-ACIA) published a white paper called "5G for Non-Public Networks for Industrial Scenarios Published" [23] in March 2019, which defines four NPN deployment scenarios, as shown in figure 3-161:

(1) The NPN in Scenario 1 is a completely isolated network compared to PLMN. The physically independent NPN ensures data security isolation and a local closed loop.
(2) In Scenario 2, the NPN and PLMN share RAN nodes but have independent Core Network equipment. The NPN and PLMN share RAN node infrastructure and even RAN cell resources, which can reduce deployment sites and costs.
(3) In Scenario 3, the NPN and PLMN share RAN nodes and control plane, and only have independent user planes. The control plane is controlled and managed by the operator. The independent subscriber plane can be placed

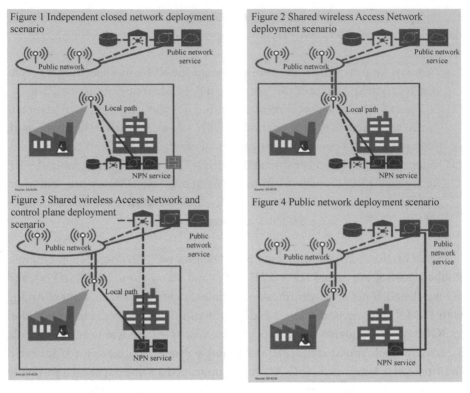

Figure 3-161 Four deployment scenarios defined by 5G-ACIA

close to the enterprise, which has a shorter path to the subscriber plane and also ensures the security isolation of enterprise data.

(4) In Scenario 4, the NPN and PLMN fully share the RAN node, control plane, and user plane. Both control plane and user plane are uniformly controlled and managed by the operator, and the mobile operator can use specific DNNs to support NPN services.

3.12.3.3 3GPP-defined deployment model

The 3GPP Rel-16 protocol supports two NPN deployment modes: PNI-NPN and SNPN, as shown in figure 3-162.

1. Public network integration deployment model

In the public network integration deployment model, the NPN depends on the PLMN for deployment. The NPN can be a specific slice in the PLMN or can be implemented by a specific DNN. In this case, the UE has the contracted data of the PLMN, which includes the contracted S-NSSAI used in the NPN. Depending

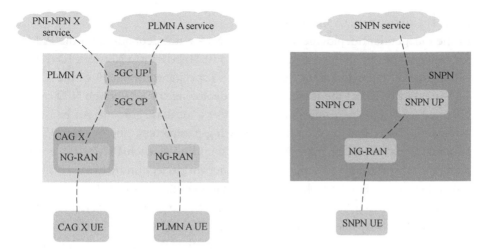

Figure 3-162 Two NPN deployment modes

on the agreement between the mobile operator and the NPN service provider, this NPN can be deployed only in the range of certain TAs. When the UE moves out of this range, the network may instruct the UE not to continue using the current S-NSSAI in the new registration area.

On this basis, the concept of Closed Access Group (CAG) can be used in order to prevent unauthorized UEs from accessing the NPN cell. A CAG represents a group of users that can access a certain CAG cell. It is represented by a CAG ID, which is unique within a PLMN. A CAG cell can broadcast one or more CAG IDs; it can also broadcast the user-readable network name corresponding to the CAG ID for manual selection by the user.

The contract data of the UE adds a list of allowed CAGs and an indication of whether access to the CAG cell is allowed only. During the UE registration process, the AMF determines whether the UE is allowed to access this CAG cell based on the UE's contract data.

After the UE is successfully registered, during the switching process in the connected state, the source-side NG-RAN ensures that the UE is not switched to a target CAG cell to which it is not allowed to access based on the list of allowed CAGs; if the mobility restriction parameters include an indication that only CAG cells are allowed to be accessed, the UE cannot be switched to a non-CAG cell either.

When the list of allowed access CAGs in the contract changes, the AMF will release the current NAS signaling connection of the UE if the CAG ID of the CAG cell currently accessed by the UE has been removed from the list of allowed access CAGs.

2. Independent deployment mode

In the independent deployment mode, SNPNs are not dependent on the PLMN for deployment. An SNPN is represented by a PLMN ID and a Network ID (NID). Where the MCC portion of the PLMN ID has a specific value of 999, or the PLMN ID is the PLMN ID of the mobile operator deploying the SNPN; the NID can be obtained from a global uniform assignment or a local assignment. An NG-RAN node with SNPN may broadcast the following parameters: one or more PLMN IDs, and a list of one or more NIDs corresponding to each PLMN ID, a user-readable network name to support manual network selection by the user, and an indication to prohibit access by other UEs that do not support SNPN.

SNPN-supported UEs may be configured with one or more SNPN contract data. An SNPN-capable UE may be set to SNPN access mode. In SNPN access mode, this UE selects and registers to the SNPN: the UE reads the list of available PLMN IDs and NIDs from the broadcast and selects the SNPN identified by the PLMN ID and NID that matches its own sign-up. If this UE has multiple signed SNPNs, the selection priority of these SNPNs depends on the UE implementation.

A UE with SNPN capability that is not set to SNPN access mode will perform a PLMN selection process to access the PLMN network. A UE may also have the ability to access both SNPN and PLMN, for example, having two Uu ports to access SNPN and PLMN networks respectively. In this case, the SNPN mode acts only on the Uu interface that has access to the SNPN.

After the UE registers and establishes a PDU session in the SNPN, that UE can establish an IPSec tunnel to the PLMN's N3IWF through the user side of the PDU session in the SNPN and register to the PLMN through that IPSec tunnel, and vice versa. The interworking adopts the architecture and mechanism of the UE access to 5GC through untrusted non-3GPP and is mentioned in other sections of this book.

3.12.3.4 Summary

5G-ACIA describes four NPN deployment scenarios and demonstrates the deployment relationship between NPN and PLMN. To support the vertical industry-oriented deployment of 5G systems, 3GPP defines two deployment modes, public network integration, and stand-alone deployment, and describes the enhanced features of 5G systems in terms of network identification, terminal configuration, and network selection in both modes. Based on these features, the 5G system provides terminal access control to support the vertical industry's demand for independent resources and realizes public-private network service interoperability.

References

1. 3GPP. Technical Specification 23.501: System Architecture for the 5G System; Stage 2 [Internet]. 2019 Jun [cited 2019 Dec 10]. Available from: http://www.3gpp.org/ftp/Specs/archive/23_series/23.501/

2. 3GPP. Technical Specification 23.502: Procedures for the 5G System; Stage 2 [Internet]. 2019 Jun [cited 2019 Dec 10]. Available from: http://www.3gpp.org/ftp/Specs/archive/23_series/23.502/

3. 3GPP. Technical Specification 23.503: Policy and Charging Control Framework for the 5G System; Stage 2 [Internet]. 2019 Jun [cited 2019 Dec 10]. Available from: http://www.3gpp.org/ftp/Specs/archive/23_ series/23.503/

4. 3GPP. Technical Specification 38.300: NR; NR and NG-RAN Overall Description [Internet]. 2019 Jun [cited 2019 Dec 10]. Available from: http://www.3gpp.org/ftp/Specs/archive/38_series/38.300/

5. 3GPP. Technical Specification 33.501: Security architecture and procedures for 5G system [Internet]. 2019 Jun [cited 2019 Dec 10]. Available from: http://www.3gpp.org/ftp/Specs/archive/33_series/33.501/

6. 3GPP. Technical Specification 23.401: General Packet Radio Service (GPRS) enhancements for Evolved Universal Terrestrial Radio Access Network (E-UTRAN) access [Internet]. 2019 Jun [cited 2019 Dec 10]. Available from: http://www.3gpp.org/ftp/Specs/archive/23_series/23.401/

7. 3GPP. Technical Specification 23.216: Single Radio Voice Call Continuity (SRVCC); Stage 2 [Internet]. 2019 Jun [cited 2019 Dec 10]. Available from: http://www.3gpp.org/ftp/Specs/archive/23_series/23.216/

8. NGMN. 5G WHITE PAPER [Internet]. 2015 Feb [cited 2019 Dec 10]. Available from: https://www.ngmn.org/work-program/5g-white-paper.html

9. 3GPP. Technical Specification 29.500: 5G System; Technical Realization of Service Based Architecture; Stage 3 [Internet]. 2019 Jun [cited 2019 Dec 10]. Available from: http://www.3gpp.org/ftp/Specs/archive/ 29_series/ 29.500/

10. IETF. RFC 8259: The JavaScript Object Notation (JSON) Data Interchange Format [Internet]. 2017 Dec [cited 2019 Dec 10]. Available from: https://www.rfc-editor.org/rfc/rfc8259.txt

11. IETF. RFC 7540: Hypertext Transfer Protocol Version 2 (HTTP/2) [Internet]. 2015 May [cited 2019 Dec 10]. Available from: https://www.rfc-editor.org/rfc/rfc7540.txt

12. ETSI. Mobile Edge Computing (MEC); Framework and Reference Architecture [Internet]. 2016 Mar [cited 2019 Dec 10]. Available from: https://www.etsi.org/deliver/etsi_gs/MEC/001_099/003/01.01.01_ 60/ gs_MEC003v 010101p.pdf

13. 3GPP. Technical Specification 23.288: Architecture enhancements for 5G System (5GS) to support network data analytics services [Internet]. 2019 Jun [cited 2019 Dec 10]. Available from: http://www.3gpp. org/ftp/Specs/ archive/23_series/23.288/

14. 3GPP. Technical Specification 28.552: Management and orchestration; 5G performance measurements [Internet]. 2019 Jun [cited 2019 Dec 10]. Available from: http://www.3gpp.org/ftp/Specs/archive/28_ series/28.552/

15. 3GPP. Technical Specification 37.320: Universal Terrestrial Radio Access (UTRA) and Evolved Universal Terrestrial Radio Access (E-UTRA); Radio measurement collection for Minimization of Drive Tests (MDT); Overall description; Stage 2 [Internet]. 2018 July [cited 2019 Dec 10]. Available from: http://www.3gpp.org/ftp/ Specs/archive/37_series/37.320/

16. 3GPP. Technical Specification 23.285: Architecture enhancements for V2X services [Internet]. 2019 Mar [cited 2019 Dec 10]. Available from: http://www.3gpp.org/ftp/ Specs/archive/23_series/23.285/

17. 3GPP. Technical Specification 23.287: Architecture enhancements for 5G System (5GS) to support Vehicle-to-Everything (V2X) services [Internet]. 2019 Aug [cited 2019 Dec 10]. Available from: http://www. 3gpp.org/ftp/ Specs/archive/23_ series/23.287/

18. 3GPP. Technical Specification 22.186: Enhancement of 3GPP support for V2X scenarios; Stage 1 [Internet]. 2019 Aug [cited 2019 Dec 10]. Available from: http:// www.3gpp.org/ftp/Specs/archive/22_series/ 22.186/

19. IEEE. 802.1CB-2017—IEEE Standard for Local and metropolitan area networks— Frame Replication and Elimination for Reliability [Internet]. 2017 Sept 28. Available from: https://ieeexplore.ieee.org/document/8091139

20. 3GPP. Technical Specification 23.316: Wireless and wireline convergence access support for the 5G System (5GS) [Internet]. 2019 Jun [cited 2019 Dec 10]. Available from: http://www.3gpp.org/ftp/Specs/ archive/23_series/ 23.316/

21. GSMA. FS.13 Network Equipment Security Assurance Scheme-Overview; NESAS Release 1 [Internet]. 2019 Oct [cited 2019 Dec 10]. Available from: https://www.gsma. com/security/wp-content/uploads/2019/11/ FS.13-NESAS-Overview-v1.0.pdf

22. 3GPP. Technical Specification 22.261: Service requirements for next generation new services and markets; Stage 1 [Internet]. 2019 Jun [cited 2019 Dec 10]. Available from: http://www.3gpp.org/ftp/ Specs/archive/22_ series/22.261/

23. 5G-ACIA. 5G Non-Public Networks for Industrial Scenarios [Internet]. 2019 July [cited 2019 Dec 10]. Available from: https://www.5g-acia.org/fileadmin/5G-ACIA/ Publikationen/5G-ACIA_White_Paper_5G_for_Non-Public_Networks_for_ Industrial_Scenarios/WP_5G_NPN_2019_01.pdf

CHAPTER 4

An Outlook on the Future Evolution of 5G Network Architecture

4.1 3GPP version release pace

3GPP operates on a project basis in the form of a series of releases, which can be interpreted as a work plan, with the work carried out in strict accordance with the predefined planning timeline. Each release is divided into three stages where Stage 1 focuses on the discussion of scenario requirements, Stage 2 is responsible for the design of the network framework, and Stage 3 is responsible for the design of specific protocols and interfaces.

In addition, typically, 3GPP is not only responsible for researching new communication technologies but also for maintaining previous or current communication technologies. For example, in 2017, 3GPP was conducting research on 5G Rel-16 technology while further refining 5G Rel-15 technology. Information about the specific pacing of Rel-14 to Rel-16 versions is shown in figure 4-1.

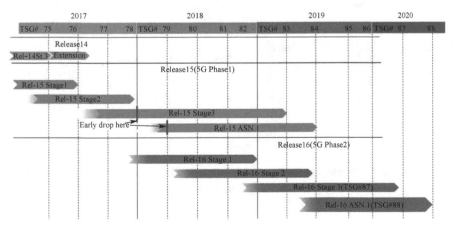

Figure 4-1 3GPP Rel-14 to Rel-16 release pace information

See table 4-1 for the functional freeze dates and protocol stabilization dates for Rel-14 through Rel-16.

Table 4-1 Rel-14 to Rel-16 functional freeze dates and protocol stabilization dates

Version number	Status	Function freeze date (End of Stage 3)	End date (Agreement stable)
Release 17	Open	TBC	TBC
Release 16	Open	2020-03-20 (SA#87)	2020-06-19 (SA#88)
Release 15	Frozen	2019-03-22 (SA#83)	2019-06-07 (SA#84)
Release 14	Frozen	2017-03-10 (SA#75)	2017-06-09 (SA#76)

Note: For details, please refer to https://www.3gpp.org/specifications/releases.

Based on the conclusion of the 3GPP SA#84 meeting, the freeze dates for Rel-17 Stage 1, Stage 2, and Stage 3 are tentatively set at 2019 Q4, 2020 Q3, and 2021 Q2, as shown in figure 4-2.

For details, please refer to https://www.3gpp.org/ftp/tsg_sa/TSG_SA/TSGS_84 /Docs/SP-190529.zip.

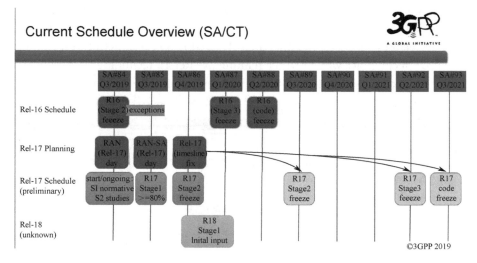

Figure 4-2 5G architecture for Rel-17 standard pacing (based on 3GPP SA#84 meeting conclusions)

4.2 Continuous evolution of the 5G network architecture

4.2.1 Introduction

The Rel-15 version of the 5G network architecture focuses on eMBB scenarios and provides initial support for other scenarios, while the Rel-16 version complements and improves support for URLLC and mIoT scenarios. The development of the Rel-17 standard has also begun. This version is a continuous evolution of the 5G network architecture, focusing on both the continuous enhancement of existing features and introduction of new features. The ongoing enhancement of existing features includes mobile edge computing, network intelligence, and a series of vertical industry oriented feature enhancements. Each version of 3GPP will introduce more features, so only the more typical ones will be discussed here.

4.2.2 Edge computing

Considering that the EC deployment scenario is one of the key enabling technologies in 5G system architecture for reducing service delay, enhancing service experience, and reducing backbone load, both 3GPP SA2 and SA6 will continue to enhance support for EC deployment scenarios in the Rel-17 phase [1, 2].

From the SA2 system architecture level, EC enhancements are supported in the following five main areas:

(1) Addressing discovery for locally deployed application servers: after the application server is deployed to the EC environment, the UE needs to discover the application server with the best route based on its current location.
(2) Open localization capability: certain rapidly changing network state information needs to be provided to locally deployed application servers through a fast open mechanism.
(3) Support for seamless application migration: in the UE migration process, in order to ensure the shortest end-to-end transmission path, the user-plane anchor UPF and the application server need to perform synchronized migration. In this migration process, identifying how to ensure message transmission without packet loss and disorder is a problem that needs to be solved.
(4) Service chain support: EC systems need to support service chain functionality internally, which is similar to the traditional Gi-LAN service chain. The main need is to investigate how to define a unified mechanism to support N6-LAN service chain functions in both EC and non-EC environments in 5G networks.
(5) Application-unaware anchor migration: anchor migration is often accompanied by a change in a UE's IP address, which in turn leads to a re-

establishment of the TCP connection and, thus, service interruption. This problem focuses on how to achieve anchor migration without application awareness.

Meanwhile, 3GPP SA6 also plans to define the service platform architecture and middleware required for deploying mobile edge computing from the service platform level, including application discovery, authentication, capability opening, and related API definition to support edge deployment.

4.2.3 Smart networks

The introduction of data analysis capabilities in mobile network architectures is a completely new topic. Rel-16 has successfully defined the basic smart network framework and standardized technical solutions for some Use Cases, but further refinements and enhancements are needed [3].

1. SMART NETWORK ARCHITECTURE ENHANCEMENTS

The Rel-16 NWDAF-based smart network framework is shown in figure 4-3. Logically, Rel-16 NWDAF can be divided into three modules: data storage, training platform, and inference platform. These modules can be logically unified, but since the training platform requires the use of a specialized hardware platform such as GPU, the following two issues need to be investigated further from the perspective of practical deployment:

(1) Whether Rel-16 NWDAF needs to do further network function division, and how to do so?
(2) If such division is not needed, is it necessary to study NWDAF deployment suggestions?

2. SUPPORT FOR ROAMING SCENARIOS

Due to its early release, Rel-16 does not consider roaming scenarios of smart networks, but along with increasing globalization, both international roaming across borders and domestic roaming across regions are very common, as shown in figure 4-4.

In general, the behaviors of roaming users and non-roaming users are rather different, as are the requirements for the network, so it is necessary to study smart network issues further in roaming scenarios.

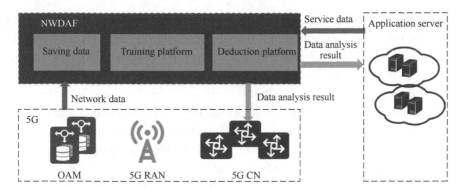

Figure 4-3 Rel-16 smart network framework based on NWDAF

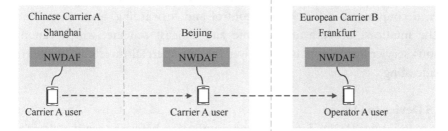

Figure 4-4 Roaming scenarios for smart networks

In addition to the issues above, we will continue to investigate ways of improving the efficiency of data collection, UPF data reporting, QoS prediction, feedback of quasi-real-time data analysis results, smart cities based on the smart 5G network, etc.

4.2.4 Multicast broadcasting

The distribution of homogeneous content through multicast broadcasting can significantly improve the efficiency of resource utilization, especially the valuable air interface frequency resources. Multicast broadcasting capability has been available in mobile networks since 3G when the 3GPP defined the Multicast/Broadcast Multimedia Subsystem (MBMS) for 3G networks and later the enhanced MBMS (eMBMS) for 4G networks. However, due to various factors such as complex technical solutions, weak demand, and incomplete terminal industry chains, multicast broadcasting has not seen large-scale commercial use in the 3G and 4G era.

The situation has changed significantly in the 5G era. Using the eMBB scenario as an example, live video broadcasting has become a mainstream application in the current mobile Internet, occupying a very large proportion of data traffic. In addition, vehicular networking and vertical industry-oriented campus networks have also put forward clear demands for multicast broadcasting. Therefore, providing the multicast broadcast function in 5G networks once again becomes the standardization focus for 3GPP.

The goal of the 5G multicast broadcasting topic [4] is to identify and evaluate potential enhancements in the system architecture to provide multicast services for different vertical services. Its research includes defining the overall framework of multicast broadcasting and the functional splitting to support different types of multicast broadcasting services with different requirements, such as V2X, IoT, TV, and live video, and to support flexible network deployment.

As for the technologies, compared with 3G and 4G, 5G multicast broadcasting may adopt design principles such as decoupling of the service layer and network layer, decoupling of network layer control and forwarding, and normalization of the multicast broadcasting scheme and unicast scheme, which can meet various service requirements and lower the commercial threshold of multicast broadcasting.

4.2.5 Device to Device

Device to Device (D2D), i.e., direct communication between terminal devices, is a capability that was supported in the 4G era. Its application scenarios are divided into two main categories: commercial and public security.

D2D features supported by 3GPP standards in the 4G era include the following:

(1) mutual detection of terminal devices
(2) D2D communication between two terminal devices
(3) group communications consisting of multiple terminals
(4) communication without network coverage
(5) terminal devices that do not have network coverage access the network through relay devices

5G will first continue to support the features already implemented in 4G, in addition to introducing some new scenarios and features [5], such as:

(1) QoS enhancement for D2D links,
(2) the two terminal devices communicating through a relay device, and
(3) terminal devices within the network coverage area being connected to the network through relay devices.

4.2.6 Other vertical industry-oriented continuous enhancements

The 3GPP Rel-16 release Vertical LAN topic focused on the enhancement of the network architecture for vertical industry requirements. This was a good start, and due to the early release time and new needs from vertical industries, 3GPP decided to continue the work towards Rel-17 release and then develop improved vertical industry-oriented standard features [6–8].

In terms of 5G LAN, more networking scenarios will be considered, for example, for terminal devices that access the 5G network through UE trunking, 5G LAN services can also be provided; in terms of 5G LAN group management, devices from different operators and roaming user devices will be supported. In addition, the selection of data exchange nodes will be optimized further to improve the efficiency of data exchange.

For TSN, Rel-17's research focus includes support for TSN multiple clock domains as well as uplink clock synchronization, and support for UE-to-UE delay-sensitive communication. It will also support new requirements for audio and video programming (e.g., concert and other scenarios) and medical scenarios that are being studied by SA1.

For NPN, Rel-17 will continue to optimize and simplify the 5GS for vertical industries, including

(1) access authentication and authorization for independent NPN deployments provided by third-party entities,
(2) online contracting and management of NPN terminal equipment,
(3) guaranteed service continuity through mixed PLMN and NPN deployment,
(4) network architecture customization and optimization for NPN deployment scenarios,
(5) support for open capabilities required for NPN,
(6) support NG-RAN sharing between SNPN and PLMN,
(7) support for emergency services in SNPN

4.2.7 Summary

After two versions of standardization (Rel-15 and Rel-16), the 5G network architecture is stable enough to meet the needs of large-scale commercial deployment and has already been officially commercialized by several carriers. The subsequent 5G network architecture will continue to introduce new and enhanced features, especially for vertical industries. While continuing 2G, 3G, and 4G to provide communication for people and things, it will also provide communication services for thousands of industries and will eventually build a smart world where everything is connected.

References

1. 3GPP. SP-190185: Study of the enhancement of support for Edge Computing in 5GC [Internet]. 2019 Mar [cited 2019 Dec 10]. Available from: https://www.3gpp.org/ftp/ tsg_sa/TSG_SA/TSGS_83/Docs/SP-190185. zip

2. 3GPP. SP-190065: Study of Application Architecture for enabling Edge Applications [Internet]. 2019 Mar [cited 2019 Dec 10]. Available from: https://www.3gpp.org/ftp/ tsg_sa/TSG_SA/TSGS_83/Docs/ SP-190065.zip

3. 3GPP. SP-190451: Study of Enablers for Network Automation for 5G—phase 2 [Internet]. 2019 May [cited 2019 Dec 10]. Available from: https://www.3gpp.org/ftp/ tsg_sa/TSG_SA/TSGS_84/Docs/SP-190451.zip

4. 3GPP. SP-190442: Study of architectural enhancements for 5G multicast-broadcast services [Internet]. 2019 May [cited 2019 Dec 10]. Available from: https://www.3gpp. org/ftp/tsg_sa/TSG_SA/TSGS_84/Docs/ SP-190442.zip

5. 3GPP. SP-190443: Study of System enhancement for Proximity based Services in 5GS [Internet]. 2019 May [cited 2019 Dec 10]. Available from: https://www.3gpp.org/ftp/ tsg_sa/TSG_SA/TSGS_84/Docs/SP-190443.zip

6. 3GPP. SP-190454: Study of the enhancement of support for 5G LAN-type service [Internet]. 2019 May [cited 2019 Dec 10]. Available from: https://www.3gpp.org/ftp/ tsg_sa/TSG_SA/TSGS_84/Docs/SP-190454.zip

7. 3GPP. SP-190630: Study of enhanced support of Industrial IoT—TSC/URLLC enhancements [Internet]. 2019 Aug [cited 2019 Dec 10]. Available from: https:// www.3gpp.org/ftp/tsg_sa/TSG_SA/TSGS_ 85/Docs/ SP-190630.zip

8. 3GPP. SP-190453: Study of enhanced support of Non-Public Networks [Internet]. 2019 May [cited 2019 Dec 10]. Available from: https://www.3gpp.org/ftp/tsg_sa/ TSG_SA/TSGS_84/Docs/SP-190453.zip

Abbreviations

3GPP	3rd Generation Partnership Project
5GC	5G Core
5G-GUTI	5G Globally Unique Temporary Identifier
5G-RG	5G Residential Gateway
5QI	5G QoS Identifier
5GS	5G System
AAA	Authentication, Authorization, and Accounting
AF	Application Function
AI	Artificial Intelligence
AMBR	Aggregate Maximum Bit Rate
AMF	Access and Mobility Management Function
AN	Access Network
ANDSP	Access Network Discovery & Selection Policy
API	Application Programming Interface
APT	Advanced Persistent Threat
AR	Augmented Reality
ARP	Allocation and Retention Priority
AS	Access Stratum
ATM	Asynchronous Transfer Mode
AUSF	Authentication Server Function
BP	Branching Point
BSS	Base Station Subsystem

CAG	Closed Access Group
CAGR	Compound Annual Growth Rate
CHF	Charging Function
C-IoT	Cellular IoT
CPE	Customer Premise Equipment
CS	Circuit Switched
CT	Communication Technology
CUPS	Control and User Plane Separation
DC	Dual Connectivity
DDoS	Distributed Denial of Service
DHCP	Dynamic Host Configuration Protocol
DN	Data Network
DN-AAA	Data Network Authentication Authorisation Accounting
DNN	Data Network Name
DRB	Data Radio Bearer
DSLAM	Digital Subscriber Line Access Multiplexer
DS-TT	Device-side TSN Translator
EC	Edge Computing
eMBB	enhanced Mobile Broadband
eNA	enabler of Network Automation for 5G
ETSI	European Telecommunications Standards Institute
FN-RG	Fixed Network Residential Gateway
FQDN	Fully Qualified Domain Name
FRER	Frame Replication and Elimination for Reliability
FS	Fault Supervision
FTP	File Transfer Protocol
FWA	Fixed Wireless Access
GAN	Generic Access Network
GFBR	Guaranteed Flow Bit Rate
GMSC	Gateway Mobile Switching Center
GPRS	General Packet Radio Service

gPTP	generalized Precision Time Protocol
GSM	Global System for Mobile Communications
GTP	GPRS Tunneling Protocol
GTP-U	GPRS Tunneling Protocol-User Plane
GW-C	GateWay-Control plane
GW-U	GateWay-User plane
HLR	Home Location Register
HR	Home Routed
HPLMN	Home PLMN
ICT	Information and Communications Technology
IGMP	Internet Group Management Protocol
IMS	IP Multimedia Subsystem
IoT	Internet of Things
IPSec	Internet Protocol Security
ISG	Industry Specification Group
IT	Information Technology
ITU-R	International Telecommunications Union-Radio communications Sector
I-WLAN	Interworking WLAN
LAN	Local Area Network
LADN	Local Area Data Network
LTE	Long Term Evolution
MAC	Media Access Control
MCC	Mobile Country Code
MDBV	Maximum Data Burst Volume
MEC	Mobile Edge Computing
MFBR	Maximum Flow Bit Rate
MICO	Mobile Initiated Connection Only
mIoT	massive Internet of Things
MME	Mobility Management Entity
mMTC	massive Machine Type Communications
MOS	Mean Opinion Score

MS	Mobile Station
MSC	Mobile Switching Center
MWC	Mobile World Congress
N3IWF	Non-3GPP Interworking Function
NAS	Non-Access-Stratum
NEF	Network Exposure Function
NF	Network Function
NFV	Network Functions Virtualization
NGMN	Next Generation Mobile Networks
NG-RAN	Next Generation Radio Access Network
NID	Network ID
NPN	Non-Public Network
NR	New Radio
NRF	Network Repository Function
NSI	Network Slice Instance
NSSAI	Network Slice Selection Assistance Information
NWDAF	Network Data Analytics Function
NW-TT	Network TSN Translator
OAM	Operations, Administration, and Management
OLT	Optical Line Terminal
ONF	Open Networking Foundation
OPEX	Operational Expenditure
PCC	Policy and Charging Control
PCF	Policy Control Function
PCO	Protocol Configuration Option
PDB	Packet Delay Budget
PDU	Protocol Data Unit
PEI	Permanent Equipment Identifier
PER	Packet Error Rate
P-GW	Packet data network GateWay
PLMN	Public Land Mobile Network

PM	Performance Measurement
PNI-NPN	Public Network Integrated NPN
PPP	Point-to-Point Protocol
PPPoE	Point-to-Point Protocol over Ethernet
PS	Packet Switched
PSM	Power Saving Mode
QCI	QoS Class Identifier
QFI	QoS Flow Identifier
QoS	Quality of Service
RAN	Radio Access Network
RDS	Reliable Data Service
RFSP	RAT/Frequency Selection Priority
RG	Residential Gateway
RNA	RAN Notification Area
RQA	Reflective QoS Attribution
RQC	Reflective QoS Control
RQI	Reflective QoS Indication
RSN	Redundancy Sequence Number
RSRP	Reference Signal Received Power
RSRQ	Reference Signal Received Quality
SCP	Service Communication Proxy
SDN	Software Defined Network
SEPP	Security Edge Protection Proxy
S-GW	Serving Gateway
SINR	Signal to Interference plus Noise Ratio
SM	Session Management
SMF	Session Management Function
SNPN	Stand-alone NPN
S-NSSAI	Single Network Slice Selection Assistance Information
SRVCC	Single Radio Voice Call Continuity
SSC	Session and Service Continuity

SST	Slice/Service Type
SUCI	Subscription Concealed Identifier
SUPI	Subscription Permanent Identifier
TA	Tracking Area
TAC	Type Allocation Code
TCP	Transmission Control Protocol
TLS	Transport Layer Security
TNAP	Trusted Non-3GPP Access Point
TNGF	Trusted Non-3GPP Gateway Function
TSC	Time Sensitive Communication
TSCAI	Time Sensitive Communication Assistance Information
TSN	Time Sensitive Network
UDM	Unified Data Management
UDR	Unified Data Repository
UDSF	Unstructured Data Storage Function
UE	User Equipment
UL CL	Uplink Classifier
UMA	Universal Mobile Access
UPF	User Plane Function
URLLC	Ultra Reliable and Low Latency Communication
URSP	UE Route Selection Policy
V2X	Vehicle to Everything
Vo5G	Voice over 5G
VPLMN	Visited PLMN
VR	Virtual Reality
W-AGF	Wireline Access Gateway Function
W-CP	Wireline access Control Plane protocol
WDR	Wide Dynamic Range
WLAN	Wireless Local Area Network
WLANSP	WLAN Selection Policy
W-UP	Wireline access User Plane protocol

Index

A

Access and Mobility Management Function
(AMF), 30–32, 37–40, 42, 43, 45, 46, 47,
49–61, 67, 68, 69, 70, 71, 72, 73, 76, 83, 86,
87, 88, 93, 94, 99, 101, 102, 103, 104, 106,
108, 109, 110, 111, 121, 122, 123, 125, 126,
127, 128, 136, 137, 139, 142, 143, 149, 152,
155, 157, 158, 159, 161, 163, 168, 169, 170,
172, 176, 178, 179, 180, 181, 182, 183, 184,
185, 186, 187, 188, 189, 190, 192, 193, 210,
213, 221, 225, 228, 235, 239, 240, 241, 242,
243, 269

Access Network (AN), 1, 4, 5, 25, 26, 30, 32,
37, 38, 40, 42, 46, 48, 51, 58, 59, 62, 63, 68,
69, 70, 71, 72, 74, 75, 76, 77, 79, 80, 83, 88,
90, 93, 95, 96, 97, 120, 125, 126, 130, 133,
143, 179, 182, 215, 216, 220, 222, 223, 224,
225, 226, 227, 228, 229, 230, 231, 232, 234,
235, 236, 237, 238, 239, 240, 243, 244, 245,
250, 251, 252

Access Network Discovery & Selection Policy
(ANDSP), 90, 93

Access Stratum (AS), 124, 125, 178, 180, 200,
201, 205, 207, 208, 209, 210, 222, 223, 225,
227, 237

Aggregate Maximum Bit Rate (AMBR), 68,
73, 82, 95, 107

Allowed Area or Non-Allowed Area, 48

Application Function (AF), 11, 29, 31, 35, 36,
83, 87, 88, 89, 94, 96, 131, 134, 135, 136,
137, 139, 140, 141, 142, 143, 145, 146, 148,
149, 151, 155, 156, 158, 159, 161, 162, 163,

165, 167, 169, 170, 172, 173, 183, 186, 187,
189, 190, 193, 200

Application Programming Interface (API), 8,
112, 119, 175, 182, 183, 187, 189, 191, 276

application stratum, 252

Artificial Intelligence (AI), 14, 22, 138

AS Release Assistance Information (AS RAI),
178

Asynchronous Transmission Mode (ATM),
229

Augmented Reality (AR), vii, 10, 14, 15, 16,
211, 212, 219

Authentication, Authorization, and
Accounting (AAA), 68, 71, 72, 73, 245

Authentication Server Function (AUSF), 30,
31, 51

availability after DDN failure, 190

average window, 80, 81, 96

B

Binding Support Function (BSF), 87

Branching Point (BP), 131, 133, 134

"Break-before-Make", 64

C

Capital Expenditure (CAPEX), 21

Cellular V2X (C-V2X), 194, 195, 197, 203,
211

Change of SUPI-PEI association, 190

Changing Function (CHF), 31, 83, 88, 95

Circuit Switched (CS) Domain, 1–5, 98, 100

Closed Access Group (CAG), 269

CM-CONNECTED state, 38, 39, 184
CM-IDLE state, 38, 180
CN type change, 191
communication failure, 190
Communication Pattern (CP), 161, 162, 188,
    233, 234, 235, 236, 237, 239, 240, 253
Compound Annual Growth Rate (CAGR),
    17
Configured NSSAI, 122, 123, 124
Connected Time enhancement, 183
Connection Management (CM), 38, 39, 180,
    184
Connection Management Connected
    Radio Resource Control Inactive (CM-
    CONNECTED RRC Inactive), 38
Connection Resume process, 180
Connection Suspend process, 180, 181
Control and User Plane Separation (CUPS),
    6, 7, 14, 22, 26
Control Plane C-IoT 5GS Optimization, 176
Core Network, 1, 2, 3, 4, 5, 8, 9, 10, 11, 25,
    29, 30, 31, 34, 35, 48, 52, 53, 54, 58, 62, 63,
    74, 80, 81, 82, 103, 112, 113, 120, 123, 138,
    139, 143, 144, 148, 149, 158, 174, 175, 191,
    192, 193, 195, 198, 199, 201, 215, 216, 217,
    218, 221, 222, 223, 226, 227, 228, 229, 230,
    231, 232, 233, 235, 237, 238, 239, 240, 241,
    251, 252, 254, 256, 267
Core Network type restriction, 48

D
Data Network (DN), 2, 5, 30, 50, 62, 64, 66,
    68, 71, 72, 73, 96, 131, 133, 136, 215, 247
Data Network Name (DNN), 30, 65, 67, 68,
    87, 90, 91, 92, 93, 96, 111, 127, 135, 136,
    137, 143, 149, 150, 151, 164, 182, 213, 214,
    247, 268
Data Radio Bearer (DRB), 53, 57, 223, 226
Dedicated Core Network (DCN), 175
deregistration, 37, 38, 40, 59, 60, 61, 224, 225,
    226, 231, 234, 235, 236, 240, 241
Deterministic Network (DetNet), 212
Device to Device (D2D), 203, 278
Device Trigger, 174
Digital Subscriber Line Access Multiplexer
    (DSLAM), 23, 26, 232
DN Authorization Data, 73

Downlink Data Delivery Status, 186, 191
Dual Connectivity (DC), 213, 214
Dual Registration (DR), 104, 111

E
Early Data Transmission (EDT), 175, 178
Edge Computing (EC), viii, xiii, 14, 130, 131,
    133, 134, 135, 136, 137, 152, 275, 276
enabler of Network Automation for 5G
    (eNA), 138, 152, 173
enhanced MBMS (eMBMS), 277
Enhanced Mobile Broadband (eMBB), 14,
    15, 24, 119, 120, 275, 277
Ethernet, 23, 62, 63, 65, 66, 73, 77, 87, 96,
    176, 177, 224, 229, 234
Ethernet PDU session type, 65, 224
European Telecommunications Standards
    Institute (ETSI), 20, 131, 132, 257
Evolved Packet Core (EPC), 5, 8, 9, 48, 98,
    104, 112, 175, 183, 184, 186, 188, 191, 192,
    193, 210
Expected UE Behavior, 178
Extended Access Barring (EAB), 174
Extensible Authentication Protocol (EAP-
    5G), 222, 223, 225, 227, 229, 230, 231, 234,
    235, 236, 237, 239
external data network (Internet/Intranet), 3

F
5G Alliance for Connected Industries and
    Automation (5G-ACIA), 267, 268, 270
5G Industrial Internet of Things, 17
5G Residential Gateway (5G-RG), 219, 220,
    232, 233, 234, 235, 236, 237, 239, 240, 241,
    242, 243, 244, 245, 246, 247, 248, 249
Fixed Network Residential Gateway (FN-
    RG), 220, 232, 238, 239, 240, 241, 242, 243,
    244
fixed wireless access, 218, 219, 220, 249
flat architecture, 5
Flow Table, 22
Forbidden Area, 48
Frame Replication and Elimination for
    Reliability (FRER), 213
Fully Qualified Domain Name (FQDN), 90,
    114, 152
function decoupling, 10

G

Gateway GPRS Support Node (GGSN), 3–4
GBR QoS Flow (QoS Flows that require a
  guaranteed flow bit rate), 78, 79, 82, 106,
  198
generalized Precision Time Protocol (gPTP),
  264, 265
Generic Access Network (GAN), 26
Generic Routing Encapsulation (GRE), 223
Global System for Mobile Communications
  (GSM), 1, 4

H

Handover, 40, 52, 55
Home Location Register (HLR), 3
home PLMN (HPLMN), 33, 90, 93, 94, 122,
  123, 128, 129, 190
home stratum, 252
H-SFN (Hyper-System Frame Number), 184
Hypertext Transfer Protocol (HTTP), 8, 29,
  118

I

implicit detach timer, 43
Industry Specification Group (ISG), 20, 131
initial registration, 46, 51, 67, 123, 210
Integrated Services Digital Network (ISDN),
  2
Intermediate UPF (I-UPF), 63, 133, 215, 216
Internet Key Exchange Version 2 (IKEV2),
  222, 223, 227
Internet Protocol Security (IPSec), 221, 222,
  223, 224, 225, 226, 227, 228, 229, 230, 231,
  234, 236, 237, 238, 240, 251, 256, 270
Interworking WLAN (I-WLAN), 26
IP Multimedia Subsystem (IMS), 2, 4, 5, 30,
  64, 97, 98, 99, 100, 103

L

LAN (local area networks), 12, 262, 263, 275,
  279
Legacy Wireline access Control Plane
  protocol (L-W-CP), 239, 240
Line ID, 247, 248
Local Area Data Network (LADN), 50, 72,
  92, 131, 136, 137

location reporting, 43, 190
Long Term Evolution (LTE), 5, 8, 9, 26, 97,
  155, 194, 195, 196, 197, 203, 204

M

"Make-before-Break", 64
massive Internet of Things (mIoT), xi, 14, 24,
  119, 120, 166, 275
massive Machine Type Communications
  (mMTC), 174, 194
Match Fields, 22
Maximum Data Burst Volume (MDBV), 80,
  96, 198
Mean Opinion Score (MOS), 147, 148, 150
Media Gateway (MGW), 4
Mobile Edge Computing (MEC), 131, 132
Mobile Equipment (ME), 210
Mobile Initiated Connection Only (MICO),
  39, 42, 44, 46, 47, 61, 183, 184, 185, 186
mobile reachable timer, 43, 46
Mobile Station (MS), 3
Mobile Switching Center (MSC), 2, 3, 4, 101,
  102
Mobile World Congress (MWC), 17
Mobility Management Entity (MME), 5, 26,
  31, 101, 102, 103, 104, 106, 108, 109, 110,
  111, 174, 192
Mobility Pattern (MP), 47, 49, 50, 158
mobility registration update, 40, 47, 54
mobility restriction, 47, 48, 269
mobility restriction list, 48
Monitoring Key, 95
MSC server, 4
Multicast/Broadcast Multimedia Subsystem
  (MBMS), 277
Multimedia Broadcast Multicast Service, 4

N

NAS Release Assistance Information (NAS
  RAI), 178
NB-IoT (Narrow Band Internet of Things),
  174, 175, 178, 191, 192, 193, 194
NB-IoT UE Priority, 178, 192
Network Address Translation (NAT), 249
network architecture, viii, ix, xi, xii, xiii, 1, 2,
  3, 4, 5, 6, 7, 8, 9, 10, 11, 13, 14, 18, 20, 22,

23, 24, 25, 26, 27, 29, 30, 31, 32, 34, 36, 83,
    84, 101, 111, 113, 119, 131, 138, 139, 173,
    174, 175, 191, 194, 195, 275, 276, 279
Network Data Analysis Function (NWDAF),
    11, 31, 83, 138, 139, 140, 141, 142, 143,
    144, 145, 146, 147, 148, 149, 150, 151, 152,
    153, 154, 155, 156, 157, 158, 159, 160, 161,
    162, 163, 164, 165, 166, 167, 169, 170, 171,
    172, 173, 202, 276, 277
Network Exposure Function (NEF), 11, 30,
    35, 36, 83, 88, 89, 131, 134, 139, 141, 145,
    146, 148, 170, 182, 183, 186, 187, 188, 189,
    191, 193, 202
Network Functions (NF), 8, 10, 20, 112, 113,
    114, 115, 116, 117, 118, 149, 151, 152, 153,
    154, 157, 158, 160, 161, 165, 172, 190, 215
Network Functions Virtualization (NFV),
    20, 119
Network ID (NID), 270
Network Repository Function (NRF), 30,
    87, 113, 114, 115, 117, 118, 122, 125, 126,
    127, 128, 129, 139, 141, 152, 154, 155, 157,
    158, 210
Network Slice Instance (NSI), 114, 123, 126,
    127, 128
Network Slice Selection Function (NSSF), 30,
    122, 125, 126, 127, 128
Network Slice Selection Policy (NSSP), 123
New Radio (NR), 8, 9, 26, 97, 191, 203, 204
Next Generation Mobile Networks (NGMN),
    111
Next Generation Radio Access Network
    (NG-RAN), 30, 39, 40, 42, 44, 45, 46, 47,
    48, 58, 59, 74, 76, 77, 81, 82, 88, 97, 98, 99,
    100, 101, 102, 104, 106, 108, 110, 141, 178,
    179, 180, 181, 182, 184, 185, 192, 195, 198,
    199, 200, 201, 214, 215, 216, 217, 221, 225,
    226, 227, 228, 232, 269, 270, 279
NF instance, 113, 114, 115
NF service, 8, 113, 115, 117
NF service operation, 113
Nnf type, 113
Non-3GPP Interworking Function (N3IWF),
    93, 221, 222, 223, 225, 226, 227, 228, 270
non-access-stratum (NAS), 30, 38, 48, 51, 58,
    60, 73, 104, 124, 175, 176, 178, 179, 182,

183, 184, 185, 187, 190, 192, 221, 222, 223,
    225, 226, 227, 228, 229, 232, 233, 234, 236,
    237, 238, 239, 241, 242, 254, 269
Non-GBR QoS Flow (QoS Flows that do not
    require a guaranteed flow bit rate), 78, 79
Non-IP Data Delivery (NIDD), 182, 183
Non-Public Network (NPN), 266, 267, 268,
    269, 270, 279
notification control, 81, 82, 96, 97, 106, 199,
    201
NSI ID, 126, 127, 128

O
OLT (Optical Line Terminal), 232
OpenFlow controller, 21–22
OpenFlow switch, 21–22
Open Networking Foundation (ONF), 21
Operation, Administration, and
    Management (OAM), 114, 138, 139, 141,
    142, 143, 148, 149, 152, 154, 155, 157, 158,
    202
Operational Expenditure (OPEX), 21, 23

P
Packet Data Convergence Protocol (PDCP),
    214
Packet data network Gateway (P-GW), 5
Packet Data Network (PDN), 5, 32, 62–64,
    68, 72, 107, 108, 109
Packet Delay Budget (PDB), 80, 198, 215,
    216, 217
Packet Detection Rule (PDR), 75, 76, 83
Packet Error Rate (PER), 80, 198
Packet Switched (PS) Domain, 1–5
Paging Occasion, 40
Paging Proceed Flag (PPF), 43
Path Switch Request, 53, 181
periodic registration update, 40, 43, 185
periodic registration update timer, 40, 43,
    185
periodic RNAU timer, 46
Per Packet ProSe Priority (PPPP), 203
Per Packet ProSe Reliability (PPPR), 203
PH (Paging H-SFN), 184
Point-to-Point Protocol over Ethernet
    (PPPoE), 234, 235, 239, 240

Point-to-Point Protocol (PPP), 229, 234
Policy and Charging Control (PCC), 60, 76, 77, 89, 94, 95, 96, 106, 107, 168, 201
Policy and Charging Rules Function (PCRF), 5
Policy Control Function (PCF), 30, 35, 36, 48, 49, 51, 55, 57, 60, 61, 68, 69, 70, 71, 73, 76, 77, 82, 83, 86, 87, 88, 89, 90, 91, 93, 94, 95, 96, 97, 106, 107, 126, 134, 139, 142, 143, 148, 149, 151, 155, 166, 168, 170, 197, 200, 201, 204, 210, 211, 247
Power Saving Mode (PSM), 174, 184
Presence Reporting Area (PRA), 95
Priority Level, 80
Protocol Configuration Option (PCO), 107
Protocol Data Unit (PDU), 51, 52, 53, 54, 55, 56, 57, 58, 59, 60, 61, 62, 63, 64, 65, 66, 67, 69, 70, 71, 72, 73, 74, 75, 77, 82, 86, 87, 89, 90, 91, 92, 93, 94, 95, 96, 103, 106, 107, 108, 109, 121, 122, 123, 126, 127, 128, 136, 137, 149, 156, 170, 176, 177, 178, 179, 180, 181, 182, 183, 192, 197, 201, 213, 214, 217, 224, 242, 243, 245, 246, 247, 248, 249, 270
Protocol Data Unit Session (PDU session), 51, 52, 53, 54, 55, 56, 57, 58, 59, 60, 61, 62, 63, 64, 65, 66, 67, 69, 70, 71, 72, 73, 74, 75, 77, 82, 87, 89, 90, 91, 92, 93, 94, 95, 96, 106, 109, 122, 123, 126, 127, 128, 136, 137, 149, 156, 170, 176, 177, 178, 179, 180, 182, 183, 192, 201, 213, 214, 217, 224, 242, 243, 245, 246, 247, 248, 249, 270
Proximity Service (ProSe), 197, 203
PTW (Paging Time Window), 184
Public Data Network (PDN), 2
Public Network Integrated NPN (PNI-NPN), 267, 268
Public Switched Telephone Network (PSTN), 2

Q

QoS Class Identifier (QCI), 79, 100
QoS Enforcement Rule (QER), 75, 76, 83
QoS Flow Identifier (QFI), 74, 76, 77, 78, 149, 224
QoS Notification Control (QNC), 106, 199, 201

QoS Profile, 75, 76, 77, 200, 201
QoS Rule, 76, 77, 78, 83, 96
Quality of Service (QoS), xiii, 5, 6, 9, 27, 29, 30, 53, 54, 57, 58, 62, 68, 69, 70, 72, 74, 75, 76, 77, 78, 79, 80, 81, 82, 83, 94, 95, 96, 97, 99, 100, 106, 107, 108, 120, 142, 147, 148, 149, 150, 151, 172, 173, 174, 175, 192, 195, 198, 199, 200, 201, 202, 203, 204, 205, 206, 209, 214, 215, 216, 217, 218, 220, 226, 262, 265, 277, 278

R

Radio Network Controller (RNC), 5
RAN Notification Area (RNA), 40, 45, 46
RAN Notification Area Update (RNAU), 40, 46
RAT/Frequency Selection Priority (RFSP), 94
RAT restriction, 48
Reflective QoS Attribute (RQA), 77, 79, 81
Reflective QoS Control (RQC), 77, 78, 83
Reflective QoS Indication (RQI), 77, 78
registration accept message, 46
registration area (RA), 37, 39, 42, 43, 44, 47, 49, 50, 54, 68, 123, 126, 136, 158, 269
Resource Type, 80
roaming status, 6, 190
Robust Header Compression (ROHC), 177
RSN (Redundancy Sequence Number), 214

S

Security Edge Protection Proxy (SEPP), 33, 256
service architecture, xiii
Service architecture, 8
service area restriction, 49, 93, 94, 97
Service Capability Exposure Function (SCEF), 6, 191
Service Data Flow (SDF), 77, 78
Service Experience (SE), 147, 148, 150, 151
Service Gap, 175
Service Level Agreement (SLA), 148, 152, 200, 216
service request, 37, 38, 40, 48, 51, 56, 57, 58, 72, 133, 137, 179, 222, 227, 229, 232, 234, 237, 238, 242, 243
service stratum, 252

Serving Gateway (S-GW), 5, 26
Serving GPRS Support Node (SGSN), 3–4
Session Aggregate Maximum Bit Rate
    (Session-AMBR), 68, 73, 82, 95, 107
Session and Service Continuity mode (SSC
    mode), 63, 90
Session Management Function (SMF), 30,
    31, 32, 51, 53, 54, 55, 56, 57, 58, 59, 60, 61,
    65, 66, 67, 68, 69, 70, 71, 72, 73, 75, 76, 77,
    78, 81, 82, 83, 86, 87, 88, 94, 95, 96, 100,
    104, 106, 107, 108, 109, 110, 111, 121, 127,
    128, 129, 131, 133, 134, 135, 136, 137, 139,
    142, 143, 149, 152, 165, 166, 168, 170, 172,
    177, 179, 180, 182, 183, 186, 187, 189, 191,
    192, 193, 201, 213, 214, 215, 216, 217, 247,
    248, 264, 266
Single Network Slice Selection Assistance
    Information (S-NSSAI), 65, 67, 68, 91, 92,
    108, 114, 120, 121, 122, 123, 124, 125, 126,
    127, 128, 135, 143, 149, 150, 151, 182, 213,
    214, 268, 269
Single Radio Voice Call Continuity (SRVCC),
    98, 100, 101, 102, 103
Single Registration (SR), 104
Slice Differentiator (SD), 120
Slice/Service Type (SST), 120
Software Defined Network (SDN), 14, 20,
    22, 119
Stand-alone NPN (SNPN), 267, 268, 270, 279
Subscription Permanent Identifier (SUPI),
    96, 143, 159, 160, 163, 164, 165, 168, 172,
    190, 254

T
TCP (Transmission Control Protocol), 118,
    222, 276
3rd Generation Partnership Project (3GPP),
    4, 5, 6, 11, 14, 23, 24, 25, 26, 29, 35, 36, 37,
    38, 39, 42, 45, 48, 51, 58, 66, 74, 81, 87, 91,
    97, 100, 102, 103, 105, 106, 108, 109, 118,
    120, 123, 131, 138, 147, 151, 154, 158, 160,
    162, 163, 164, 170, 172, 173, 198, 199, 206,
    215, 217, 218, 219, 220, 221, 222, 223, 224,
    225, 226, 227, 228, 231, 232, 234, 235, 236,
    239, 240, 243, 245, 246, 249, 250, 256, 257,
    258, 261, 262, 264, 266, 267, 268, 270, 273,
    274, 275, 276, 277, 278, 279

Time Sensitive Communication Assistance
    Information (TSCAI), 265, 266
Time Sensitive Communication (TSC), 265,
    266
Time Sensitive Network (TSN), 212, 213,
    264, 265, 266, 279
Time to Market (TTM), 10
TLS (Transport Layer Security), 118, 251, 256
Tracking Area (TA), 42
transport stratum, 252
trusted Non-3GPP Access Point (TNAP),
    228, 230, 244
trusted Non-3GPP Gateway (TNGF), 228,
    245

U
UE Aggregate Maximum Bit Rate (UE-
    AMBR), 82
UE communication, 162, 163, 164, 165, 167,
    170
UE Configuration Update (UCU), 48, 49,
    123, 136, 211
UE Reachability, 186
UE Route Selection Policy (URSP), 65, 66,
    67, 68, 90, 91, 92, 123, 126, 127, 128, 213
Ultra Reliable and Low Latency
    Communication (URLLC), 14, 24, 119,
    120, 211, 212, 214, 216, 217, 275
Unicast Link Profile, 207
Unified Data Management (UDM), 30, 35,
    49, 51, 60, 61, 65, 68, 69, 71, 104, 109, 110,
    111, 125, 139, 158, 166, 170, 185, 189, 190,
    191, 193
Unified Data Repository (UDR), 30, 31, 35,
    36, 83, 88, 89, 184
Universal Integrated Circuit Card (UICC),
    210
Universal Mobile Access (UMA), 26
Unstructured Data Storage Function
    (UDSF), 30, 31, 35, 36
unstructured PDU session type, 65, 224
Uplink Classifier (UL CL), 104, 131, 133, 134
User Equipment (UE), 6, 29, 30, 32, 33, 36,
    37, 38, 39, 40, 42, 43, 44, 45, 46, 47, 48, 49,
    50, 51, 52, 53, 54, 55, 56, 57, 58, 59, 60, 61,
    62, 63, 64, 65, 66, 67, 68, 69, 70, 71, 72, 73,
    74, 75, 76, 77, 78, 79, 80, 81, 82, 83, 86, 87,

88, 89, 90, 91, 92, 93, 94, 96, 97, 98, 99,
100, 101, 102, 103, 104, 105, 106, 107, 108,
109, 110, 111, 121, 122, 123, 124, 125, 126,
127, 128, 133, 135, 136, 137, 142, 143, 148,
149, 150, 154, 155, 156, 158, 159, 160, 161,
162, 163, 164, 165, 166, 167, 168, 169, 170,
171, 172, 173, 174, 175, 176, 177, 178, 179,
180, 181, 182, 183, 184, 185, 186, 187, 188,
189, 190, 191, 192, 193, 195, 197, 198, 202,
203, 204, 205, 206, 207, 208, 209, 210, 211,
213, 214, 215, 217, 219, 220, 221, 222, 223,
224, 225, 226, 227, 228, 229, 230, 231, 232,
235, 236, 237, 238, 242, 243, 244, 245, 250,
254, 262, 263, 264, 265, 266, 267, 268, 269,
270, 275, 276, 279
User Plane C-IoT 5GS Optimization, 176
User Plane Function (UPF), 29, 30, 32, 38,
45, 54, 55, 56, 57, 58, 60, 62, 63, 64, 65, 66,
67, 68, 70, 71, 72, 75, 76, 77, 78, 79, 80, 81,
82, 83, 88, 104, 106, 108, 121, 131, 133,
135, 136, 137, 140, 142, 143, 149, 152, 168,
179, 180, 182, 186, 187, 188, 213, 214, 215,
216, 217, 239, 245, 247, 248, 249, 262, 263,
264, 265, 275, 277

V

Virtual Reality (VR), vii, 10, 14, 15, 16, 212,
219
visiting PLMN (VPLMN), 33, 90, 93, 94, 122,
128, 129, 192
Visitor Location Register (VLR), 3
Voice over LTE (VoLTE), 97, 98, 100, 101,
103

W

Wideband Code Division Multiple Access
(WCDMA), 4
Wide Dynamic Range (WDR), 18
Wireline access Control Plane protocol (W-
CP), 233, 234, 235, 236, 237, 239, 240
Wireline Access Gateway Function
(W-AGF), 232, 234, 235, 237, 238, 239,
240, 241, 242, 243
WLAN Selection Policy (WLANSP), 93

Tan Shiyong, a graduate of Tsinghua University with a master's degree in Pattern Recognition and Intelligent Systems, has been in R&D in the mobile communication industry for 18 years. His experience spans the standard formulation of 3G, 4G, and 5G mobile network architectures. Since 2013, he has dedicated his efforts to 5G network architecture research and standard work, serving as the head of the Network Technology Group of China's IMT-2020 (5G) Promotion Group. Currently, he leads Huawei's 5G Network Architecture Standard Research Team, which has been at the forefront of 5G network architecture standard development.

Ni Hui, a graduate of Tianjin University with a PhD in Computer Science and Technology, is a Principal Engineer of Standard Research at Huawei Technologies Co., Ltd. and an expert in wireless network architecture technology. She has been instrumental in standardizing technologies like 5G network architecture, network slicing, URLLC, edge computing, and control and forwarding separation in 4G networks.

Zhang Wanqiang, a graduate of the Beijing University of Posts and Telecommunications with a master's degree in Communication and Information Systems, is a standards expert at Huawei Technologies Co., Ltd. He has contributed significantly to the standardization research of 4G/5G mobile network system architecture, mobility management, NB-IoT system architecture, and more.

Wu Xiaobo, a graduate of Northeastern University with a master's degree in Computer Software, is a Principal Engineer of Standard Research at Huawei Technologies Co., Ltd. He has participated in the research direction of 4/5G voice and 5G network intelligence and led the standardization work of 3GPP SA2 5G intelligence.

Hu Huadong, a graduate of Tianjin University with a PhD in Electrical Theory and New Technology, is a standards research expert at Huawei Technologies Co., Ltd. He has been involved in and led research directions and standardization work related to 3G PCC, 4G SAE & CUPS, and 5G service architecture. Hu is a senior standards expert in wireless network architecture and a 5G security standards expert at Huawei Technologies Co., Ltd.

Together, this team of seasoned professionals offers unrivaled insights into the world of 5G. Their combined expertise has shaped the standards of 5G technology and driven its evolution.